Violence Against Women in Early Modern Performance

Also By Kim Solga

PERFORMANCE AND THE CITY *(edited with D.J. Hopkins and Shelley Orr)*

Violence Against Women in Early Modern Performance

Invisible Acts

Kim Solga

 UNIVERSITY OF WINCHESTER LIBRARY

© Kim Solga 2009

All rights reserved. No reproduction, copy or transmission of this publication may be made without written permission.

No portion of this publication may be reproduced, copied or transmitted save with written permission or in accordance with the provisions of the Copyright, Designs and Patents Act 1988, or under the terms of any licence permitting limited copying issued by the Copyright Licensing Agency, Saffron House, 6–10 Kirby Street, London EC1N 8TS.

Any person who does any unauthorized act in relation to this publication may be liable to criminal prosecution and civil claims for damages.

The author has asserted her right to be identified as the author of this work in accordance with the Copyright, Designs and Patents Act 1988.

First published 2009 by
PALGRAVE MACMILLAN

Palgrave Macmillan in the UK is an imprint of Macmillan Publishers Limited, registered in England, company number 785998, of Houndmills, Basingstoke, Hampshire RG21 6XS.

Palgrave Macmillan in the US is a division of St Martin's Press LLC, 175 Fifth Avenue, New York, NY 10010.

Palgrave Macmillan is the global academic imprint of the above companies and has companies and representatives throughout the world.

Palgrave® and Macmillan® are registered trademarks in the United States, the United Kingdom, Europe and other countries.

ISBN: 978–0–230–21954–0 hardback

This book is printed on paper suitable for recycling and made from fully managed and sustained forest sources. Logging, pulping and manufacturing processes are expected to conform to the environmental regulations of the country of origin.

A catalogue record for this book is available from the British Library.

A catalog record for this book is available from the Library of Congress.

10 9 8 7 6 5 4 3 2 1
18 17 16 15 14 13 12 11 10 09

Printed and bound in Great Britain by
CPI Antony Rowe, Chippenham and Eastbourne

For my mom and dad

Contents

List of Illustrations   ix

Acknowledgements   x

1 **Encounters with the Missing: From the Invisible Act to In/visible Acts**   1
     The Duchess's ghost (Stratford, 2006)   4
     Violence as invisibility in early modern England   7
     Performance as witness (to the missing)   12
     The violations of theory   19
     In/visible acts   26

2 **Rape's Metatheatrical Return: Rehearsing Sexual Violence in *Titus Andronicus***   29
     The politics of 'hue and cry'   32
     Messing up (in) the theatre of sexual trauma   42
     Spectacles of absence   50

3 **The Punitive Scene and the Performance of Salvation: Violence, the Flesh, and the Word**   63
     Staging grace against the odds   63
     Correction, salvation, and the companionate negotiation   67
     Making a spectacle of salvation   78
     Body doubles, Babel's voices   85

4 **Witness to Despair: The Martyr of Malfi's Ghost**   98
     Witnesses and ghosts   98
     The Duchess dies a martyr   103
     Bodies, acts, texts   106
     Performing despair   111
     Ghosts in the machine   117
     Model witness   129

5 **The Architecture of the Act: Renovating Beatrice Joanna's Closet**   141
     Rape, space, and reception desire   142
     Looking at Beatrice Joanna   146

Crimes of architecture	151
Renovating Beatrice Joanna's closet: the Cheek by Jowl *Changeling*	159
Afterword	**176**
Notes	180
Bibliography	187
Index	205

List of Illustrations

1. Brian Cox as Titus and Sonia Ritter as Lavinia in Deborah Warner's 1987 production of *Titus Andronicus* at the Swan Theatre. Photo by permission of the Joe Cocks Studio Collection, Shakespeare Birthplace Trust. 55
2. Michael Maloney as Frankford and Saskia Reeves as Anne in Katie Mitchell's 1992 production of *A Woman Killed With Kindness* for the RSC. Photo by Leah Gordon, with her kind permission. 87
3. *The Duchess of Malfi*, 2006. From left: Shane Carty as Antonio Bologna, Laura Condlln as Cariola, and Lucy Peacock as the Duchess of Malfi. Photo by David Hou. Courtesy of the Stratford Shakespeare Festival Archives. 121
4. *The Duchess of Malfi*, 2006. From left: Joyce Campion as the Old Lady, Laura Condlln as Cariola, Ronan Rees as the Boy, Shane Carty as Antonio Bologna, and Lucy Peacock as the Duchess of Malfi. Photo by David Hou. Courtesy of the Stratford Shakespeare Festival Archives. 124
5. Will Keen as Ferdinand and Janet McTeer as The Duchess (with members of the company) in Phyllida Lloyd's 2003 production of *The Duchess of Malfi* for the Royal National Theatre. Photo by Ivan Kyncl. Courtesy of Alena Melichar. 132
6. Janet McTeer as The Duchess and Lorcan Cranitch as Bosola in Phyllida Lloyd's 2003 production of *The Duchess of Malfi* for the Royal National Theatre. Photo by Ivan Kyncl. Courtesy of Alena Melichar. 138
7. Olivia Williams as Beatrice Joanna and David Collings as Vermandero in Cheek by Jowl's 2006 production of *The Changeling*. Photo by Keith Pattison, with his kind permission. 161
8. Olivia Williams as Beatrice Joanna in Cheek by Jowl's 2006 production of *The Changeling*. Photo by Keith Pattison, with his kind permission. 166

Acknowledgements

My first debt of gratitude is to my teachers. Thanks to those who were professors when this project first began, and who now are trusted colleagues and friends: Nancy Copeland, Elizabeth D. Harvey, Stephen Johnson, and Ric Knowles.

Many have read this manuscript, in whole or in part, at different stages of its development; still others have sat across tables or on porch chairs and helped me work through irritations both mild and massive. Thanks to Roberta Barker, Sylvia Hunter, M. J. Kidnie, Laura Levin, Marlis Schweitzer, Jenn Stephenson, Joanne Tompkins, and Pauline Wakeham. Without their generosity of time, energy, and spirit this book would look far different.

I have learned much from my students, who remind me every day that I have only begun to scratch the surface of the problems this book handles. Thanks to the participants in my 2005 graduate seminar, 'Architectures of Feminist Performance', and the participants in my two 2008 undergraduate senior seminars, both titled 'Performing Violence Against Women in Early Modern English Theatre'. Special thanks here to Allison Hargreaves, who is now more colleague than student and whose own work on violence against First Nations women in contemporary Canadian contexts is an inspiration to me.

For their many kindnesses, on-site support, and electronic assistance I owe a debt of gratitude to the staff at the Shakespeare Centre Library and Shakespeare Birthplace Trust, especially Helen Hargest; the staff at the Royal National Theatre archive, especially Gavin Clarke; the staff at the Stratford Shakespeare Festival archive, especially Ellen Charendoff; and the staff at Cheek by Jowl, especially Jacqui Honess-Martin.

This book could not have been completed without the Herculean efforts of my research assistants: Nadine Fladd, Karla Landells, Sarah Pesce, Elan Paulson, and Kristen Warder. Later, the help of Paula Kennedy, Steven Hall, and Penny Simmons at Palgrave proved essential.

Finally, thanks and love to my family: Linda Solga, Dieter Solga, and Jarret Hardie. And to B.M., for all the (life-saving) margaritas.

An early version of Chapter 2 was published in *Theatre Journal*, vol. 58, issue 1 (2006), pp. 53–72 (copyright The Johns Hopkins University Press); portions of Chapter 3 were published in *Contemporary Theatre*

Review, vol. 18, issue 2 (2008), pp. 146–60 (www.informaworld.com). I also acknowledge the kind financial support of the Social Sciences and Humanities Research Council of Canada (SSHRC) and the University of Western Ontario.

1
Encounters with the Missing: From the Invisible Act to In/visible Acts

> [I]t is not *presence* that appears in performance but precisely the missed encounter – the reverberations of the overlooked, the missed, the repressed, the seemingly forgotten.
>
> (Schneider, 2001, p. 104)

This is a book about violence against women around the turn of the seventeenth century in England: its pernicious erasure in cultural texts of all kinds, the negotiations of that erasure in some of the most iconic plays in English theatre history, and the rehearsal of those negotiations on our late twentieth- and twenty-first-century stages. I shape my argument around the deliberate collision of the historical and the contemporary as I try to imagine what it might mean to represent early modern experiences of violence against women on the stage in an ethical way, a feminist way, today. The moment in theatre history dominated by Shakespeare's cohort is often described as brutally spectacular. I ask: among its vivid, grotesque representations of bodies, blood, and revenge, how and why does violence against women go so spectacularly missing? What role does early modern England's heady performance culture play in the shaping of this central absence, and what legacies does it leave for theatre makers, theatre scholars, and theatregoers working on its remains now? Can we rehearse the (often indeed spectacular) disappearance of violence against women in early modern performance without reproducing it? That is, can we rehearse it with a difference, rehearse its very cultural invisibility in order to get a better purchase on how and why violence against women so often (then, and even now) comes into existence as a violent and violating evacuation?

Violence Against Women in Early Modern Performance invests in both the 'then' of the early moderns and the 'now' of contemporary Western performance, but not in order to draw simplistic transhistorical links between the two. Rather, I challenge the urge toward transhistoricity reproduced in so much contemporary early modern performance, and I explore the ramifications – especially for women – of taking early modern wonders for late modern signs. Twenty-first-century directors, producers, and audiences continue to insist on the commercial value and the cultural relevance of the turn of the seventeenth century: the world over, major regional, fringe and 'festival' theatres remain both artistically and economically dependent on the work of Shakespeare, Webster, Jonson, Middleton, and others, as both broadly universal and inherently 'present' (see Bennett, 1996, pp. 17, 20). And yet this work makes a consistent investment in the obvious oppression of women, an investment with which many contemporary audience members are not likely to be in sympathy and toward which many more may be openly hostile (Barker, 2007, p. 2). How do we square this work's enormous cultural capital with its profound distance from contemporary attitudes toward social justice and human rights?

The Elizabethan and Jacobean periods maintained what today's readers will no doubt regard as retrograde laws surrounding sexual and domestic violence, but at the same time they also enacted some significant changes to rape law, as well as engaged in fresh debates about the practice of wife-beating. These changes and debates were accompanied by subtle yet important shifts in social texts (such as household manuals and marriage sermons) as well as literary and theatrical materials, and these shifts, in turn, are phenomena for which we need to account not simply in terms of their historical status but also in terms of their dissemination in what Tracy Davis calls 'performative time' (2008), what David Román calls 'the time of the historical now' (2005, p. 15). One of my central claims in this book is that a valuable link we might indeed draw across cultures and time lies in the vicissitudes of performance itself – as it was understood to function politically and socially among the early moderns, and as its workings have been theorized over the last three decades by Herbert Blau, Joseph Roach, Peggy Phelan, Rebecca Schneider, and others. The cultural management of early modern violence against women often took plainly theatricalized forms, as spectacle intervened to remake a woman's bodily and spiritual experience of her suffering into something more socially and politically comprehensible. In turn, I wonder: how might the contemporary early modern stage develop what Roach has called a genealogy of performance in response to these theatricalized historical interventions, in order to 'attend to [...] the disparities between history as it is discursively transmitted and memory as it is publicly enacted by the bodies that bear its consequences' (Roach, 1996, p. 26)?

This book, then, is as much about contemporary performance practice as it is about the early moderns, and it is as much a theatrical and cultural history as it is a performance ethics. I shape each chapter that follows this introduction around an early modern cultural 'act' that reveals the representational history of violence against women in the period (what Catherine Belsey would call 'history at the level of the signifier' [1999, p. 5]) in an effort to develop strategies for the feminist performance of violence against women on our contemporary stages – strategies that need not be limited to the work of any historical period, but that also, crucially, strive to account for the historical specificity of any given representation of violence against women. I take a performance studies approach throughout, reading key legal, social, and religious trends from the early modern period as performative events that beg an actor's response from female victims of violence. I ask what kinds of performances these trends demand of women's bodies, and how these performances are both actualized and resisted in the dramatic work of the period. Following Diana Taylor, I define performance as 'simultaneously connoting a process, a praxis, an episteme, a mode of transmission, an accomplishment, and a means of intervening in the world' (2003, p. 15). Performance for me is a cultural doing, a historical doing, but it is also a means of cultural and historical intervention. The possibility of such intervention lies latent in the early modern dramas I examine, and it is manifested in contemporary performances of those dramas with greater or lesser degrees of success. Theatre takes place in designated performance spaces, but the effects of a broader, more socially readable theatricality – the rape victim's public 'hue and cry'; the battered woman's domestic 'performance of salvation' – can be felt in spaces both within and far beyond the theatre. The productive tension that obtains among performances of violence against women in the public square and the public–private household, the representations of those performances on the early modern stage, and the rehearsal of those representations on our own stages today forms the principal critical nexus of the book.

As I write, I deliberately blend early modern cultural and theatrical history with current debates in performance theory: about theatre's ephemeral status and its resulting power to stage the absent or missing; about performance's power to act as a witness to history and as a venue for reimagining the tenor of that history and its impact on our collective future as artists, critics, and citizens; about how theatre history and performance criticism have (or have not) been able to come to terms with the performance of violence against women on stage. I take many cues from the substantial body of feminist performance theory that has been written over the past three decades; it informs my analysis throughout, but I also take issue with some of its principal pressure points. Feminist theatre scholars have worked hard to establish how and why women's bodies on stage have been subject to a violent and violating gaze, and to develop techniques for deflecting and returning that gaze. In the wake of this important work, however, they have left largely untouched the problems that adhere to the female body in representation during moments of *literal* stage violence. This book returns to feminist performance theory at this late moment in its intellectual development in order to ask what difference such acts of violence make in the theatrical representation of women's bodies, and to build a theory and a praxis both for performing and for encountering that violence today.

What does it mean, now, to rehearse early modern violence against women as both visually stunning and culturally invisible? Can we find ways to perform the *history* of its elision – Belsey's 'history at the level of the signifier' – rather than just repeat that elision again and again for fresh spectators? These are the questions I have been taking with me, for some time now, to the theatre.

The Duchess's ghost (Stratford, 2006)

> The Ghost *should* frighten us out of our wits, for the fact is we have no wit for the Ghost. The Ghost is a meta-fact of an apprehension, an invisible event, at which we can only make mouths.
>
> (Blau, 1982, p. 213)

> Here the body [...] becomes a kind of archive.
>
> (Schneider, 2001, p. 103)

A hot early summer day in 2006. I am at the Stratford Shakespeare Festival to see Peter Hinton's new production of *The Duchess of Malfi*. I enter the dim Tom Patterson auditorium, the bleacher seating still hiding my view of the stage, and am hit by a sudden rush of air-conditioned air. I feel my body relax, my shoulders drop. Then, as I round the corner to take my place, the air in my lungs leaves me. There at the centre of the long thrust stage, watching me through sallow face and dead eyes, is the ghost of the Duchess of Malfi.

'Ghosts', writes Alice Rayner, 'hover where secrets are held in time: the secrets of what has been unspoken, unacknowledged; the secrets of the past, the secrets of the dead' (2006, p. x). They are – as we know from Herbert Blau (1982), Joseph Roach (1996), Marvin Carlson (1989), Peggy Phelan (1993; 1997) and Rebecca Schneider (2001), the 'ghost theorists' whose heritage haunts Rayner's exciting reworking of the genre – the ephemeral substance of theatre. Ghosts mark what is simultaneously visible and invisible in performance, the uncanny gap within the representation that makes us (but does not fill us with) wonder (just a tickle at the back of the brain). Ghosts stage our unknowingness, 'what the given to be seen fails to show' in the face of the spectacular (Phelan, 1993, p. 32). They are the intangible effects that 'test the limits of [our] intelligibility' (Rayner, 2006, p. xiii), pointing neither to 'presence' nor at 'absence' in some grand, metaphysical sense but toward the mundane, entirely material markers of a very palpable, very human loss. These ghosts are not reducible to, say, Hamlet's dad or even the Duchess herself, as she appears post-mortem to her husband during Webster's fifth act. These ghosts are different, undomesticated, what Rebecca Schneider might call 'performance remains'. They expose the secret question theatrical events always provoke in their simultaneous plasticity and ephemerality, the extraordinary invisibility of their labour and of so much of their rigging: 'how [do] we see what is missing' (Rayner, 2006, p. xiii)?

The figure (actor Joyce Campion) who watched me watching her as I took my seat in the Tom Patterson auditorium that June afternoon was no garden-variety ghost; she is, after all, not even really in the play. Downlit from above, her face made up to look ragged, skull-like, time-worn, the old woman on the small bench beside a still-life portrait of a death's head struck me immediately not just as the ghost of the Duchess but as the Duchess *as* ghost – as the embodied echo of the Duchess's lost body, destroyed with *Grand Guignol* fervour in Webster's Act Four only to be quite forgotten in Act Five. The Duchess herself, of course, (the Duchess as icon) has hardly been forgotten; her story is revived continually, has been the 'play of choice' for those looking to produce a Jacobean text outside the Shakespeare canon in recent performance history (Bennett, 1996, p. 86; see also Barker, 2007, p. 56). But the question of *how* we remember the Duchess – of how we read her torture and its end; of how she dies and what meanings are ascribed to her death in perpetuity – remains open, remains more a matter of the 'unspoken' and 'unacknowledged' within Webster's play than a matter of the cathartic release afforded by Bosola's eulogy for his dead mistress at the end of Act Four and the bloody Act Five in which the Duchess returns eerily to Antonio. The haunting, taunting centre-stage body with which Hinton opened his production carried, for me, the resonance of this absence and the burden of its demand. In this play that is, apparently, so overwhelmingly about the Duchess, her grief and her martyrdom, how might we see what is missing, nevertheless, of her in the frame? What would it take for us truly to encounter the Duchess's ghost?

As the performance pressed on I kept a vigilant watch for the haggard old woman from the preset. She turned up in the strangest places, and slowly I began to realize that she was not simply an echo, and not simply a neat production effect. She *was* in the play after all, and with a vengeance. She was there to stand in for many of its unseen and unspoken roles – and thus to call attention to the unseen and unspoken within the Duchess's story. Campion played an old maidservant in the Duchess's household; she played the Duchess's midwife; she played the embodiment of the myth of the House of Aragon, the one that says the ghost of an old woman appears to this family whenever a loved one dies. More significantly, she acted as a silent but omnipresent observer throughout the performance, appearing barely noticeable from scene to scene and yet making her investment in the action clear. As the Duchess's brothers entreated their sister against marriage in Act One, Campion posed on the balcony above the scene; their hypocritical commentary elicited her grim, hair-raising laughter. Later, at the Duchess's death, she quietly watched Bosola grieve. In Act Five she accompanied the Duchess across the stage during the latter's brief return, and here, finally, she openly played the role I had already assigned to her: the ghost of the Duchess's ghost, its critical double, the pregnant echo of her mistress's missing body.

After the show I found myself thinking often of Campion. Woman of many roles, simultaneously marginal yet multiple within the production, essential yet abject to the story, she made me wonder how many ghosts she might, in the end, have stood (in) for. Was her rough and weary body the ghost of the Duchess who never got to grow old, never felt her own body aging? Was she the ghost of all those endless Duchesses we've watched bathetically in stalwart past productions, Duchesses martyred only to be returned as spooky, sentimental harbingers to their families during the 'echo' scene? Or was she, indeed, the ghost of the family ghost, the spectre of twice-died death (the one who makes all their deaths into theatre, all pain into spectacle)? Rayner, like Schneider before her, teaches me that theatre's ghosts stage an encounter with elision, a 'missed encounter' that is, in fact, an encounter with missing-ness itself, as the substance of performance (2006, pp. 1–32): they remind us that performance harbours what Rayner calls the power of 'unforgetting' (2006, p. xviii), the power to awake us to the thing that has just now passed us by, before our diverted eyes. For me, this was Campion's power. She inhabited the Duchess's negative, the unseen and unheard story of her body, but she also manifested the power of the spectator as witness, one who makes a personal investment in the 'trauma of the impossible' and commits to *being there* when it passes, it if passes', commits to seeing not just 'the time of the other in representation' but also 'our own investments in the time of the other,' our own stakes in the frame (Rayner, 2006, p. 32). Ultimately, she was the ghost of each one of us in the audience, a mirror up to our own acts of watching, an interrogatory figure whose silent presence worried the very definition of spectatorship. She was the one who watched each one of us take our seats (perhaps watched us watching all along). She was the woman – both real and not, both locatable in time and space, and not – whose gaze ever trailed our perhaps too-complacent bodies and eyes with questions about how and what we were seeing, and how and what we would not, could not see.

Violence as invisibility in early modern England

The history of violence against women in early modern England is a history of just such a failure to see (Baines, 1998, p. 69). Although there are a large number of acts that could easily fall under the rubric of 'violence against women' in the late sixteenth and early seventeenth centuries (including, for example, a substantial amount of politically motivated religious violence following the Reformation, something I discuss in Chapter 4), I focus my explorations here on two of the most common, and for many women most familiar, forms: rape and the physical abuse of wives. Of course, by calling these acts violence I commit a deliberately anachronistic, ideological act of my own, for the early moderns defined a very limited array of such acts as violence proper, and saw even fewer of them as violence committed directly against a woman's body. Rape was a crime against a household, a husband or father, his goods, property and honour; domestic violence was 'reasonable correction', a form of household-ordering essential to the proper functioning of parish, county, and state. In early modern England, violence against women was a serious social and legal problem – rape mattered; unreasonable cruelty against wives also mattered – in part *because* it was defined by its very social invisibility, by its relationship to patriarchal pride and necessary forms of social control.

Feminist historians have in recent years paid growing attention to what they call the 'effacement' of violence against women, particularly rape, in Renaissance English culture (see Kahn, 1976; Stimpson, 1980; C. Williams, 1993; Baines, 1998; and Bamford, 2000).[1] The First and Second Statutes of Westminster (1275 [WI] and 1285 [WII]) enshrined rape as a property crime committed against a family or household head by aligning it with a woman's elopement or abduction (Post, 1978, p. 158). Kim Phillips has argued that the language of these statutes represented a movement away from an earlier legal model (based on the Glanvill treatise of 1187) that appeared to take women's physical suffering more directly into account as a substantive aspect of rape crime (2000, p. 129). Regardless of whether or not this precedent had once granted female victims agency in their bodily suffering, with the conflation of rape and abduction in WI and WII and the entrenchment of these positions in further statues in 1382 and 1487 rape was *de facto* equated with a wife's or daughter's potential 'seduction' by her rapist, signalling her betrayal of her household and her complicity in the crime (Burks, 1995, p. 766). In the process, her violated body became little more than a cipher for her family's place in the social economy (Phillips, 2000, p. 138). Husbands, fathers, and even the King were given proxy control over both rape and its reparations (ibid., pp. 136–7) because daughters were no longer trusted to act in 'their' best interests. Fear of women's sexual 'defection' grew (Burks, 1995, p. 766) and continued to shape the reception of rape accusations through the Tudor and Stuart periods. In an influential essay published by the London Feminist History Group in 1983, Nazife Bashar concludes that statute changes in 1555 and 1597, which made rape and abduction once again separate crimes and placed heavier emphasis on women's consent as a defining arbiter, allowed rape to be viewed as 'a crime against the person, not as a crime against property' (1983, p. 41). Yet Bashar's conclusions are optimistic: medieval laws remained in effect, and the result of the statute changes was not a sudden legal interest in rape as sexual violence against women, but rather an awkward negotiation of old prejudices and new laws.[2]

As a result of rape's socioeconomic status in the period, certain female bodies were clearly understood by the early moderns as more at risk from rape than others. Assaults on virgins yet to be bought in marriage mattered far more than assaults on older women, while women whose sexual mores were murky were often deemed consenting to the acts against them and seen thus not as victims but as accomplices.[3] Moreover, rape prosecutions, when they were attempted, demanded a tightly regulated legal performance in which a victim and her family had to work hard to establish their claim by mimicking the very specific terms through which rape was recognized as relevant to patriarchy. As Chapter 2 charts extensively, this performance relied upon a carefully coded rehearsal of rape that victims were required to play out publicly, and in which their cries of rape and claims of non-consent hinged on the success with which they made their own personal suffering disappear from the scene. Garthine Walker locates the period's effacement of sexual violence in the gendered strictures of early modern rhetorical models, which strangled women's freedom to speak about sexual matters: '[f]or women, available discourses about sex – sin and whoredom – were confessional and implicatory. Responsibility for sex, and the blame and dishonour that went with it, was feminised [...] there was no popular language of sexual non-consent upon which women could draw' (1998, pp. 5, 8). In other words, simply to speak out about rape implied a woman's loss of sexual innocence, which in turn could potentially signify her complicity in the act. Instead, rape victims talked about the act's adverse affects on their household work, their husbands' goods or their fathers' responsibilities (see also Chaytor, 1995; Hanawalt, 1998; and Gowing, 2003, esp. p. 93), turning private trauma into public experience and potentially burying, in the process, the effects of that trauma on their own minds, bodies, and spirits.

As it did with rape violence, early modern culture struggled to come to terms with the legal, social, and moral status of non-sexual domestic violence against women. Such violence was 'reasonable correction' at the turn of the seventeenth century – a male householder's *droit de maître* and the means by which he kept his little commonwealth in order. However, because the boundaries of reasonable correction were not clearly demarcated, as legal writers of the time themselves acknowledged (T.E., 1979, p. 128; see also Heale, 1974), husbands enjoyed few enforceable limits on their 'right' to subdue their wives physically (Fletcher, 1995, pp. 192–203; Somerville, 1995, pp. 89–97; F. Dolan, 1996, p. 218). Social historian Susan Dwyer Amussen (1994) has argued that a woman's community could step in where the law failed, imposing limits on excessive correction through routine neighbourhood surveillance. The spectacle of correction was always necessarily a public one, in which neighbours maintained the parallel power either to endorse a husband's management of his wife, children, and servants or to bear witness to the genuine bodily trauma 'correction' might produce and help bring the battered wife forward for justice. Yet Amussen's own evidence suggests that suits against excessive correction were often resolved by sending husband and wife home again to live quietly; authorities preferred to protect the sanctity of the household over the bodily sanctity of the wife whenever possible (ibid., pp. 78–9; see also Mendelson and Crawford, 1998, pp. 142–3).

Excessive correction, as Amussen notes, generated a troubling image of unquietness in the community, one which reflected badly on household, parish, and beyond; when reasonable correction stopped seeming reasonable, even to those trained to see nearly all forms of household violence through its lens, it threatened to reveal the extent to which early modern patriarchy relied for its image of ruling benevolence upon violence against women that was simply disavowed as such. While a husband's private brutality was of serious concern for his wife as well as his parish, more disturbing for the culture at large was the possibility that a fearful, protesting, black-and-blue wife might make a spectacle of herself and her suffering before her community or the courts, drawing attention to the broader failure of 'reasonable correction' as an ordering mechanism. Laura Gowing has suggested that wives who found themselves in such a position often took pains to describe their suffering in economic terms, in order to make it visible as fundamentally *dis*orderly on a larger social scale (1996, pp. 209–10). Meanwhile, writers of popular conduct texts looking for a means to square their disenchantment over reasonable correction with its persistent reality articulated a different role for the battered wife to play on the public stage: director of and star in what Chapter 3 calls her 'performance of salvation', as one who suffers patiently at the tyrant's hand and earns 'a great reward therefore' (*An Homily on the State of Matrimony*, 1992, p. 20). Even as the vast majority of conduct writers turned *against* the basic tenets of reasonable correction in their advice for husbands, many continued to provide wives with the above model for handling a cruel spouse (Cleaver, 1598, pp. 227–8; Whately, 1623, pp. 210–15; Gouge, 1976, pp. 320–1). To rise up, even against a tyrant, would be to deny wifely obedience as the most important duty a woman owed her lord(s); the patiently suffering wife could recuperate her bodily loss and resume her required duties by playing the martyr to the national good while also acting as a religious exemplar for her husband and family (Whately, 1623, p. 215). As with statute changes in rape law, then, the slow evolution of cultural attitudes against wife-beating did not necessarily translate into emergent neo-liberal beliefs about women's bodily sanctity and protection.

Rather, the immediate result of this evolution was the imminent need to stage-manage the *appearance* of ongoing household violence within the larger public sphere, a tricky performance of 'percepticide' (Taylor, 1997, pp. 119–38, esp. p. 123) rather than a smooth legal transition to modern women's rights.

My brief survey above suggests that the early modern effacement of violence against women was every inch the product of spectacle management. Specific forms of social theatre worked not only to 'disappear' acts of violence against women (Taylor, 1997), shifting those acts into a new legal or social register (property crime; family betrayal; reasonable correction; heavenly salvation), but also to normalize their disappearance, making spectacle itself the (invisible) agent of systemic atrocity. The image of a woman's suffering that threatened to overwhelm homosocial articulations of rape or 'correction' was replaced by a culturally choreographed performance of a different order, one (importantly) initiated by the female victim. A threat to the image of patriarchal benevolence became the image of benevolence itself, leaving 'the woman's humiliation and destruction', as Diana Taylor writes in another context, 'seemingly invisible to the audience and resistant to a critique' (ibid., p. 10).

Unlike in the public sphere, where the seamless performance of violence's disappearance was a high-stakes game, on the early modern stage the cultural management of sexual and domestic violence could be more easily explored and challenged. In the chapters that follow I read Shakespeare's *Titus Andronicus*, Heywood's *A Woman Killed With Kindness*, Webster's *The Duchess of Malfi* and Middleton and Rowley's *The Changeling* for their engagement with these pernicious methods of performative disappearance, and I contend that violence against women in early modern England must be understood within the framework of both the social spectacles that make its evasion possible and the play texts that simultaneously reproduce and critique the performative processes of that evasion. All of the plays I examine are tragedies; their dates span the roughly 30 years across 1600 that have proved among the most prolific in English theatrical culture. Each was popular in its moment and remains canonical today. Most importantly, each play maintains a self-conscious, non-axiomatic relationship to the performative management of violence against women on which it pivots, and each has had at least one major production in English within the last 20 years that probes that relationship critically, using acting technique, stage space, or design elements to provocative feminist ends. I am particularly interested in the work these texts, and these recent productions, do as meta-performance, the ways in which they both encode and resist the highly theatricalized strategies through which early modern culture negotiated the optics and import of women's violence. For me, the central question then becomes that of performance's own ethical status in the equation. If the turn of the seventeenth century in England witnessed the collusion of violence with mimesis as it refined its strategies for dealing with the public spectre of violence against women, and witnessed those strategies in action as often at the theatre as in the public and semi-private spaces beyond the theatre, can performance *also* become a witness to the very processes of cultural blinding it appears to facilitate (Taylor, 1997, pp. 125–6)?

Performance as witness (to the missing)

Performance has long been keyed, anxiously, to absence. From Plato's anxieties over the potential hidden meanings of an actor's impersonations (what it might really mean to 'play' the Philosopher King) to Artaud's desire to make what lies beneath the hegemony of representation spring forth in the loaded space between performer and spectator (and thus to absent representation altogether), theories of performance are always, on a deep level, concerned with the fact that mimesis conceals more than it reveals, stages the lost and missing within the image of plenitude it presents sometimes as truth and sometimes as the failure of human access to truth. Absence, as Peggy Phelan first argued in 1993, is both performance's figure and its ground; performance's import lives more in its silences than in its languages, and in those silences its power to connect human beings across sexual, racial, social, and ideological differences also appears (Margolin, 2008). And yet the relationship between performance and absence is in no way simple, nor ideologically neutral, nor historically fixed. Responding to Phelan's controversial argument that performance's ephemerality *is* its defining ontology, Rebecca Schneider writes: 'in privileging an understanding of performance as a refusal to remain, do we ignore other ways of knowing, other modes of remembering, that might be situated precisely in the ways in which performance remains, *but remains differently?*' (2001, p. 101, my emphasis). Within the last 15 years, Phelan's argument has blended with Schneider's nuance as the relationship between performance and its 'missing' has taken on a new, politicized urgency, becoming the subject of serious consideration by scholars interested in exploring performance's power to act as a witness – to the processes of psycho-social subject formation that govern our interactions with one another in the public sphere; to political upheaval; to national or community trauma; and to the buried histories that lie behind all three. Performance is often understood as an activist genre, an art form produced in tandem with a live, omnipresent audience ready to be inspired to flee the theatre and make social change. But this new investment in performance-as-witness is not reducible to older imaginings of theatre as social activism; rather, it is 'fast becoming indicative of an epistemology grounded in loss and mourning' (Rayner, 2006, p. 101), indebted to the catastrophe of 11 September 2001 but also not defined exclusively by it. In much of the recent critical work on this subject, performance 'enable[s] a reimagination of history' (Román, 2005, p. 3); it becomes an act of historical surrogation, an intervention into 'the cavities created by loss through death or other forms of departure' (Roach, 1996, p. 2), a site at which audiences 'constitute themselves as a provisional public [...] in order to rethink their relationship to the official national performance of mourning and patriotism' (Román, 2005, p. 278). Performance witnesses what remains unvisualized and unacknowledged in the increasingly spectacular late modern public sphere and its increasingly haphazard remembrance of its own historical legacies (Schneider, 2001 and 2009). It 'becomes itself through messy and eruptive reappearance', a fading-into-memory that nevertheless bears with it a disruptive, insistent materiality (2001, pp. 103, 106). At their most radical and their most hopeful (J. Dolan, 2005; Román, 1998), these theories of performance-as-witness galvanize around performance's power to stage an encounter with what has gone missing in public culture, and to make space for audience members to imagine their relationship to both that culture and its acts of disappearance anew.[4]

This contemporary movement toward understanding performance as, specifically, the performance of what *we* miss in 'just looking' without exploring critically our acts of seeing (Taylor, 1997, p. xii) begins with Peggy Phelan's groundbreaking work on disappearance as a feminist representational strategy in *Unmarked* (1993). In that book and in her later *Mourning Sex* (1997) Phelan writes performance as the cultural site at which we are best able to witness our own subjectivity as a measure of a basic ocular lack, our inability to see ourselves save through the purloined gaze of another (1993, p. 23). Thinking performance theory through Lacanian psychoanalysis, she argues that this lack-in-seeing is a condition we cannot fully recognize but which we nevertheless must attempt to approach in order for an intersubjective ethics, a legitimate encounter with the other in his or her proper alterity, to become possible. Phelan subsequently theorizes performance as a fundamentally non-reproducible event, and therefore as one in which all those bodies – gendered bodies, raced bodies, traumatized bodies – absented or disappeared through normative processes of representation (processes that depend on the making-same of the other) may come into blurred half-focus as the by-products of a pervasive psychic and cultural disavowal, of elision as a strategy of individual as well as national subject-formation.

All work on theatrical absence and performative witness – including my own here – is indebted to some degree to the psychoanalytic frameworks of Freud and Lacan, as each posits a psychic and social afterlife beyond lived, articulated consciousness (what we don't see, and how we don't see it), and as each understands the power and the potential dangers that may attend the return to Symbolic space of what Lacan calls 'the Real'. Phelan, for example, does not advocate the unproblematic return of the missing to representation; instead, she envisions this return as uncanny, as 'unmarked', or as what Schneider might call the 'messy and eruptive reappearance' of representation's lost bodies:

> The unmarked is not spatial; nor is it temporal; it is not metaphorical; nor is it literal. It is a configuration of subjectivity which exceeds, even while informing, both the gaze and language. In the riots of sound language produces, the unmarked can be heard as silence. In the plenitude of pleasure produced by photographic vision, the unmarked can be seen as a negative. In the analysis of the means of production, the unmarked signals the un(re)productive.
> (Phelan, 1993, p. 27)

In the performance of the unmarked it is not words that speak but their echo that haunts; remains are not buried and ritualized but 'stand *as* remains rather than body' (Phelan, 1997, p. 10). Symbolic articulations, which mark 'you' and 'me' as opposing entities, binarily constituted and differentially valued, give way to Imaginary forms of intersubjective recognition that operate, awkwardly but forcefully, as 'the affective outline of what we've lost' (ibid., p. 3), what we lose when we enter the space of dualistic thinking. This outline 'cannot be assimilated into a mute object of the reader's [or viewer's] contemplation' (ibid., p. 10), but for that very reason 'might bring us closer to the bodies we want still to touch than the restored illustration can' (ibid., p. 3).

Phelan theorizes the unmarked as performance's means of staging the daunting fissure at the heart of all subjectivity; in turn, this theorizing hinges on her parallel, and now notorious, claim that performance is representation without reproduction and cannot – from an ethical perspective should not – be rehearsed or preserved in another form (1993, p. 149). Although this claim does capture something central to theatre's power to stage an act of witness, to be present to an always-receding moment in representational history, I remain concerned by the ideological implications of fetishizing disappearance as unequivocally valuable. When, and for whom, does disappearance operate as a theoretical and political advantage, and when and for whom does it promise other, larger losses (Taylor, 2003, p. 193)? In what circumstances does arguing for performance's irreproducibility become an unethical critical gesture, a nod to the opposing powers of a fixed archive and a forgetting of the power of the 'live body' as an alternate archival form (Schneider, 2001, p. 100)? Diana Taylor's work in both *Disappearing Acts* (1997) and *The Archive and the Repertoire* (2003) centres on the questions of what knowledge is permitted to appear and what knowledge is rendered invisible by normalizing cultural paradigms, and of how the former relies upon the latter for its very status *as* 'visible' knowledge. In the process of this working-through, Taylor lends a praxis to Phelan's theoretical arguments in order to remind us that 'absence', 'disappearance', and 'missing-ness' are in no way essentially benevolent or subversive terms. Insofar as they may be the condition of possibility of performance, these terms work both sides of the political coin: they can shore up vicious political regimes as quickly as they can lend power to critics of those regimes. They can activate the power of witness to atrocity and injustice, but they can also – and far more easily – commit percept... train us to 'overlook' what lies in plain sight (Taylor, 1997, p. 121) and to 'collude with the violence around [us]' (ibid., p. 123) as the institutions that commit social and political violence deploy performance tactics in order to make the seemingly obvious simply disappear. How, then, do we locate and activate the counter-discursive power in performance's fundamental engagement with the politics of absence? If performance's ontological fissures can so easily allow it to be co-opted, how and when does absence *become* performance's weapon against the hegemonies of dislocation and disappearance?

In an argument that bears the marks of, but also moves beyond, the earlier work of both Phelan and Taylor, Alice Rayner writes that performance always embeds, in the gap between the theatrical script as written (the law of language; archival representation) and the bodies that inhabit its voices (Taylor's repertoire; Phelan's unmarked), the ghosts of what neither written word nor spoken voice can fully embody. (We might argue that these ghosts are acutely palpable among the early moderns: for the sixteenth and seventeenth centuries the 'script as written' often took multiple forms, none of them decisively authoritative, while the bodies that originally gave those scripts voice are now literal ghosts for many a scholar poring over a dusty edition of the quarto or folio text. What lies between them is our radical uncertainty about what the period really looked like, who these vaunted masters really were.) History, Rayner suggests, can never appear on stage *as* history, because history 'is a ghost that rises from the gap of repetition' (Rayner, 2006, p. 42), that appears only in outline in the futile process of attempting to stage what we just don't really know. In the process of making performance we do not achieve conventional historical transmission; instead, we articulate a space between what is said and what remains unsaid and unknown about the performed experience in/as history. 'The voices carry history through the body into the world', Rayner writes. 'Such history is not a matter of documentation or information but a matter of imitation and performance that embody both material specificity and the gaps that arise in the process of imitation and repetition' (2006, p. 42). A venue for the live transmission of what can only with difficulty be understood, the theatre lets us '[work] out complex relations between remembering and forgetting'. It becomes a place where we come to realize 'not *what* has been forgotten but *that* something has been forgotten' (ibid., p. 52), to encounter denial, percepticide, and its disquieting revelation *simultaneously*. Rayner argues that it is not performance's non-reproducibility that gives it cultural currency, but rather its uncanny power to stage *its own* debts to disappearance, its inability ever fully to materialize that which it claims to represent in the first place, even in the midst of the attempt. Performance is not simply the thing that disappears, escapes the reproductive economy; it fully undermines the claims of all reproductive economies by staging a live, immediate encounter with the failure of historical knowledge, with the miss that constitutes our engagements with the events of political and social history.

Phelan's unmarked (the echo of loss staged *as echo*, as outline of absence rather than the swell of return), married with Rayner's notion of theatre as a space in which to meet the ghosts of history in the gap between hegemony and its simultaneous disarticulation, offers me a particularly powerful tool for imagining performance as a witness to what I want to call the 'in/visible' act of violence against women. The in/visible act is both/and: it enters representation as invisible, as elided within representation, but quickly becomes palpable as such, as a missed and missing story of loss *within the frame of the very performance that would complete the process of its effacement.* The in/visible act is a guerrilla performance gesture that erupts from within the spectacle of violence's elision at its most critical moment – that interrupts, messily, violence's own forgetting. It takes advantage of spectacle's power of percepticide rather than simply succumbing to it; it argues for performance's power to bear witness to its own complicity in the disappearing act, as well as to history's indebtedness to such acts for the authority of its representations. The in/visible act trades in doubles more than negatives or silences: doubled bodies, doubled voices, doubled acts, just as Joyce Campion's multiply-signifying ghost body doubled the body, voice, and actions of Lucy Peacock's Duchess in Peter Hinton's production, signifying at once the 'more than' of the Duchess's story that could not be told in that venue at that moment, as well as the 'Real' that 'was the absence of her', the continuing representation of her loss and our inability to experience its totality (Phelan, 1993, p. 12). The in/visible act, doubling spectacle's elisions while ghosting its losses, is both embedded in the process of percepticide and a cancer to its growth.

The in/visible act is the performance of violence against women as critical forgetting; it charges its witnesses to come to terms with what we've missed but also with *how* we've missed – with how we have failed to see the suffering before us, hidden in plain sight. It is a metaperformative structure: it lets us see the codes of our failed watching erupt alongside the substitute image that takes shape in the ocular gap our failure leaves behind. This is a process Kelly Oliver calls witnessing 'beyond recognition'. For Oliver, like many contemporary trauma theorists, to witness means to engage historical rupture at the level of loss, not at the level of supplementarity and sense-making; the witness is one who testifies to 'that which cannot be reported by the eyewitness, the unseen in vision and the unspoken in speech, that which is beyond recognition in history' (2001, p. 2). For performance truly to act as a witness to what has been made to disappear from history, it must frame an encounter between the missing and its audience that provokes in the latter the sense of anxiety and even vertigo attending this movement 'beyond recognition'. Roger Simon calls this feeling 'the touch of the past':

> To be touched by the past is neither a metaphor for simply being emotionally moved by another's story nor a traumatic repetition of the past reproduced and re-experienced as present. Quite differently, *the touch of the past signals a recognition of an encounter with difficult*

knowledge [...] [A]t stake in the touch of the past is the welcome given to the memories of others as a teaching – not simply in the didactic sense of an imparting of new information – but more fundamentally *as that which brings me more than I can contain.*

(2005, p. 10, my emphasis)

As I posit the in/visible act as a strategy of witness, I am not advocating the rehearsal of violence's missed trauma in order to recuperate it; such a rehearsal would serve only to soothe audiences into believing that they have, finally, been privileged to access what has so long been veiled, that now they know (what others never did know). Rather, I propose the in/visible act as a discombobulating practice that provokes in spectators the otherworldly feeling of not-quite-knowing, of radical uncertainty about one's own historical base-line.

A witness must always testify to what lies beyond ready access ('the unseen in vision and the unspoken in speech'). The witness is provoked to challenge what she believes to be true about her powers of recognition and, as Phelan might suggest, what unequal relationships vision may institute between self and other. (Oliver argues that the eye need not be a distancing organ, but can be understood as connective tissue, a means of reaching forward to another rather than capturing that other's gaze for one's own identificatory processes [2001, pp. 11–15; 1998, pp. 149–59].) A witness is also, importantly, always a witness *to history*: the process of (self-) critical witnessing challenges history's privileged forms of knowledge and representation, seeking out the 'misses' within those forms and locating the implications of history's 'miss' for a more thorough knowledge of the present. (The witness is one who understands the present historically, as Román's 'time of the historical now'.) Finally, witnessing is always a performed act. Oliver argues that 'the unseen of history is shown through performance' (2001, p. 143), that performance is the specific practice that allows us to move 'beyond recognition'. The performing body speaks (or shows, gestures) toward the missing in the historical record, and toward the 'dehumanization and objectification' (ibid., p. 98) that absence affects, while the spectator-as-witness encounters that absence as a representational process, as history at the level of the signifier, and with it a glimpse of a story unseen, untold, unheard (see also Taylor, 2003, p. 167). In order to make a meaningful connection, the witness must seek the 'trace' in the performer's testimony (Simon, 2005, p. 146); the elided past may then instantiate a community of witnesses prepared to reckon with what they do not and cannot know. The audience, for its (for our) part, 'must face [the spectacle] without being able to transform [it] into manageable meanings' (Graver, 1995, p. 63). Prepared to encounter a loss they may not be able to mourn, an audience of witnesses engages with the performance before them as a challenge to their very ways of knowing, and of 'being-in-the-world-with', others (Shildrick, 2002, p. 108).

The violations of theory

While the in/visible act as I have conceived it here (and will deploy it in future chapters) invites audiences into a deep engagement with the representational history of violence against women during a very specific moment in place and time (that of Shakespeare and his contemporaries), it is not simply a historical technique. The in/visible act does not stage past elisions as merely 'past', a safe distance from our own attitudes and practices; it also works to illuminate the centrality of similar absences in the here and now. What kinds of witnesses to violence against women are we today, especially in the academy, especially in the disciplines of theatre and performance studies? To what extent is the erasure of violence against women *still* the story of that violence in our culture at large, and how do we, or do we not, address that erasure in our work?

Feminist performance theory has since its inception in the early 1970s been concerned above all with the violence of the gaze (L. Williams, 1989, p. 189).[5] Laura Mulvey's iconic 'Visual Pleasure and Narrative Cinema' (1975) laid out the framework through which a generation of feminist scholars understood the female body in representation; in a famous and oft-quoted passage Mulvey describes the cinematic apparatus trapping the female body as object of either sadism or idolatry:

> In a world ordered by sexual imbalance, pleasure in looking has been split between active/male and passive/female. The determining male gaze projects its phantasy on to the female figure which is styled accordingly. In their traditional exhibitionist role women are simultaneously looked at and displayed, with their appearance coded for strong visual and erotic impact so that they can be said to connote *to-be-looked-at-ness*.
>
> (1975, p. 11)

'Visual Pleasure and Narrative Cinema' installed the violence of the male gaze and the phallocentric narratives that beckon it into action (ibid., p. 14) at the core concern of feminist film and performance theories through the 1970s and 1980s, but often at the expense of literal moments of violence against women on stage or screen. This is not to say that feminist theorists simply ignored all such moments (see, for example, L. Williams, 1989); it is also not to demean the value of these theorists' readings of the cultural violence done to women as a result of their object-position in a hierarchical representational matrix (de Lauretis, 1987, p. 45). Understanding representation (specular and narrative) as a form of gendered violence that both relies upon and produces binary gender difference makes all feminist work possible; I aim not to challenge the value of this work, which informs so much of what I write in this book and elsewhere. I do, however, hope to query the extent to which violence as a *trope* for feminist performance research has been allowed to overtake, and even act as a substitute for, our critical engagement with a material reality as complex, as historically and culturally specific, and in many ways as resistive as women's violence and suffering in performance.

Feminist film theory's turn toward the violence of language and the gaze led feminist theatre theorists to focus similarly, and to theorize the ways in which live female performers might return, refract, or deflect the violating gaze on stage (Diamond, 1989; 1997a). This work has been central to the legacy of feminist performance theory and continues to be urgent and relevant today. So what, then, is the problem? Can critical explorations of the violence of representation not live alongside parallel explorations of more literal forms of violence against women in performance? While there can surely be no divorcing 'literal' instances of violence in representation from the broader matter of the violence *of* representation, problems arise when representational violence becomes the ascendant term in a new epistemological binary. Jeanie Forte positions her 1992 article, 'Focus on the Body: Pain, Praxis, and Pleasure in Feminist Performance' as a deliberate response to the representational turn in feminist work, suggesting that 'the study of cultural representations alone, divorced from consideration of their relation to the practical lives of bodies, can obscure and mislead' (Bordo, qtd in Forte, p. 249). Forte builds an argument that reads women's pain and live performance as two 'circumstances in which the body is undeniable, when the body's material presence is a condition of the circumstance' (ibid., p. 251). Soon, however, her focus shifts to an analogy in which women are cast as suffering for and through 'patriarchal culture' in the broadest sense of the term; she musters the work of Elaine Scarry in order to describe women's experiences under patriarchy as a form of 'torture' (ibid., p. 252). She then comments very briefly on a pair of feminist ordeal performances before turning her attention to the representation of bodily pleasure (ibid., p. 254).

Forte looks to the suffering body in feminist performance art only to look away – from its status as a material condition, and from its historical specificity (she reads work that is broadly contemporary, but neither historicizes its representations nor defines the boundaries of the patriarchy with which she engages). The pain of her title quickly dissolves into metaphor before disappearing altogether under the sign of pleasure-as-resistance. Falling into its own blind spot, Forte's essay poses a question that nags at the larger body of feminist performance theory: why can we not seem to approach violence against women's bodies in performance without immediately diverting our eyes? So much feminist performance criticism – 'Focus on the Body' is perhaps only the plainest example – promises an engagement with women and violence in representation only to stall on a broad reading of 'patriarchal culture' as a violent and violating force whose abstract power trumps any specific instants of violence it might produce. But this looking away comes, I believe, at a steep price. When we fail to engage with historically and culturally specific representations of women's violation and suffering in our criticism, we miss the many ways in which these representations – and their effacement, then as now – both underwrite and go unnoticed within the larger bodies of performance theory and theatre history. In other words, we risk installing the erasure of women's violence, pain, and suffering in cultural and mimetic representation *as* both performance theory and theatre history (see Enders, 2004).

In *Stages of Terror: Terrorism, Ideology, and Coercion as Theatre History* (1991), Anthony Kubiak tells a 'performative history of terror' (p. 2). He argues that theatre is not merely a stage on which terror's effects are played out but also 'the locus of terror's emergence as myth, law, religion, economy, gender, class, or race' (1991, p. 5). 'Terror' does not mean 'violence' for Kubiak; on the contrary, violence is what we stage when we want to cover terror up (ibid., p. 10). Terror is related to the blind spots in thought and perception, those that contain glimpses of 'the imminence of a non-being that defines life' (ibid., p. 16), and thus 'terror' is finally Kubiak's term (*pace* Kristeva's abject) for all that we radically disavow in order to claim our subjectivity in the Symbolic, that which always threatens to return from the Real (ibid., pp. 12, 16). Terror is, for Kubiak, theatre's central absence, its great effaced: it can never appear on stage but as a forclusion (ibid., p. 10). Terror is 'real pain'; terrorism is merely 'the techniques of its production' on stage, which can only ever take us far away from apprehending true terror (ibid., p. 21).

This hierarchy of terror – with the 'real' terror of the subject at its head, and violence as a gross, unsophisticated act of elision at its base – forms Kubiak's conceptualization of terror as the motor of theatre history.

In his commitment to the blind spot at the heart of all representation *as* the history of theatrical representation, and in his call for theatre to replace stagings of 'simple' violence (1991, p. 139) with the ghosts of terror – the 'traces' of its 'disappearance' that 'play over the suspect appearance of the body' subject to terrorism (ibid., p. 161) – Kubiak is kin to both Peggy Phelan and Alice Rayner, an early voice in the attempt to formulate performance as a witness to the missing. But, at the same time, the hierarchical opposition terror/terrorism that regulates Kubiak's argument prevents him from fully recognizing what bodies, and what experiences of terror, must be disavowed in order to make this formula of terror-as-subjectivity into a theory and history of performance. To experience terror as subjectivity one must first have access to a sense of oneself as a subject, but what of all those for whom subjectivity remains an open question? What of the violence they experience in their very attempts to *claim* status as subjects? Kubiak's readings of *Othello* and *The Duchess of Malfi* are telling in this regard. In each case his reading ends at the feminist beginning: the moments in which Desdemona and the Duchess are murdered. Kubiak is interested in the terror that ghosts Othello's need for the ocular proof; Othello's murder of Desdemona becomes the 'sheer terrorism' that covers his lack (ibid., p. 63). Similarly, he focuses on Ferdinand's obsession with the dead Duchess's body *after* her systemic torture and brutal strangling; her ghostly echo in Act Five (the echo of the text's own elision of her suffering body) becomes Kubiak's trope for terror, 'the disembodied voice that once again articulates a black absence at the heart of desire, echoing through the bleak ruins of history, marking with a never present voice the forever disappearing locus of the self' (ibid., p. 69). That self is not the Duchess's, nor is it Desdemona's. It is Othello's, Ferdinand's – they are the true subjects of terror (the true subjects of theory). Violence against their women becomes the ground on which Kubiak stages theatre's debt to terror; the disappearance of that violence barely registers because, after all, it is only 'simple' violence (ibid., p. 139), and therefore not worthy of either a theory or a history of its own.

Stages of Terror tells a bold, provocative theatrical history, but by making terror a trope for both theatre history and performance theory Kubiak inadvertently closes down the possibility of reading specific moments of violence on stage historically, culturally, or corporeally. A similar problem faces Peggy Phelan at the end of *Unmarked*, as she attempts to read Angelika Festa's installation performance, *Untitled Dance (with fish and others)*. Early in her discussion Phelan tells us that the installation 'is primarily a spectacle of pain' (1993, p. 153) only to slip quietly past this comment in order to focus her attention instead on 'some of the broad claims which frame *Untitled*' including 'the lack of difference between some of Western metaphysics' tacit oppositions – birth and death, time and space, spectacle and secret' (p. 153), making Festa's performance of suffering anterior to its status as metaphor. It seems, for Phelan as for Kubiak, as though the representation of a woman in pain is unapproachable *except* as a cipher for the 'broad claims' of 'Western metaphysics' – except as sweeping evidence of both History and Theory.

I am not suggesting that we abandon theory as we seek to understand violence against women on stage; unlike David Graver, who has also been critical of both Kubiak and Phelan for their lack of engagement with the material body in pain, I do not seek to replace cruelty's 'thrall to metaphysics' (Graver, 1995, p. 49) with a discourse of the body's presentism. But I do propose that we need to be far more aware of our tendencies as critics, when faced with a crisis body, to suture its epistemological wounds and to ward off the encounter with our own un-knowingness it promises. Bodies in violence – especially those bodies whose status as violated has been denied, their wounds erased but not healed by dominant discourses – require a critical witness who attends always to the limits of his or her own ability to comprehend, lest we risk turning an already-effaced body into a model for our own cultural power. Roger Simon distinguishes between the spectator and the witness, a distinction we might extend to that between critic and witness. Spectators watch and listen actively but work hard to manage what they see and hear (Simon, 2005, p. 92); witnessing, on the other hand, is a learning practice (ibid., p. 150). It shows us the limits of spectacle management, shows us that not all knowledge can be shaped into theory. After all, as Elaine Scarry writes, 'For the person in pain, so incontestably and unnegotiably present is it that "having pain" may come to be thought of as the most vibrant example of what it is to "have certainty," while for the other person it is so elusive that "hearing about pain" may exist as the primary model of what it is "to have doubt"' (qtd in Graver, 1995, pp. 53–4).

Peggy Phelan crashes into Angelika Festa's blind, awkwardly suspended body; in response, she produces an analysis that reads like a parody of art criticism and of her own role as critic. 'Festa's wrapped body itself seems to evoke images of dead mummies and full cocoons', she writes. 'Reading the image one can say something like: the fecundity of the central image is an image of History-as-Death (the mummy) and Future-as-Unborn (the cocoon)' (1993, pp. 153–6). After framing Festa's ordeal body as a symbol for the labour of childbirth (p. 160), Phelan turns to History in a tone that embeds a mix of frustration and rage at the failure of her theoretical tools to capture the performance adequately:

> As one tries to find a way to read this suspended and yet completely controlled and confined body, images of other women tied up flood one's eyes. Images as absurdly comic as the damsel Nell tied to the railroad ties waiting for Dudley Doright to beat the clock and save her, and as harrowing as the traditional burning of martyrs and witches, coexist with more common images of women tied to white hospital beds in the name of 'curing hysteria,' force-feeding anorexics, or whatever medical malaise by which women have been painfully dominated and by which we continue to be perversely enthralled. The austere minimalism of this piece [...] *actually incites the spectator toward list-making of this type*. The lists become dizzyingly similar *until one finds it almost impossible to distinguish between Nell screaming on the railroad tracks and the hysteric screaming in the hospital.*
>
> (1993, p. 160, my emphasis)

In this remarkable paragraph, Phelan performs the process of aversion she ascribes to Festa's *Untitled Dance*, encountering the work as truly 'unmarked', as the (silent) articulation of her own critical blind-spot. Phelan is, in my reading of this reading, in no way unaware of the significance of her failure; her entire argument in *Unmarked* up to this point has been concerned with those performances, just like Festa's, that stage a critical encounter with the failure of self-seeing and that refuse to return the spectating look in order to reinforce the integrity of the watching self. The disarming, dispossessing power of Festa's performance is, in theory, wholly expected. And yet Phelan seems, at the same time, to be remarkably unprepared for what she meets in Festa's wrapped, taut, agonizing body. She appears in the pages of this analysis as a spectator reduced to the hysterics she claims to read in Festa's performance, suddenly desperate yet unable to locate a position from which to control its meanings.

Phelan's reading of *Untitled Dance* falls in the middle of the final chapter of *Unmarked* and acts as the book's climax. I believe it is not an accident that Phelan chose to end her work in this book – about what we want to see but cannot, how we miss it, and how we might, in performance, be made to see both our 'wish' as well as our 'miss' (1997, p. 12) – with a body whose suffering she cannot see, read, comprehend, or come to terms with *not* comprehending. Festa's *Untitled Dance* gestures obliquely around the periphery of a woman's ordeal performance at all the interpretations Phelan anxiously ascribes to it, and more; at its centre, though, is a suffering female body in the process of disappearing into the bandages in which Festa is wrapped so tightly as barely to appear a body. It is a performance that stages an encounter with the elision of the suffering female body as a disciplinary history, a critical history. It is a performance that stages the seductive invitation to read around it, to miss it completely. Showing us her wish as well as her miss, Phelan's spectacularly failed reading of Festa's physical suffering articulates 'the body of a woman *in pain*' (1993, p. 162) as the quintessential 'unmarked', that which Phelan's book pushes toward but that which it cannot, finally, encounter except as loss. Festa's is the body at which eyes *always* fail, the thing made to disappear so that cultural history, theatre history, and performance theory might be hung on its bones.

How can we get past the troping of violence against women for both performance theory and theatre history? How do we move beyond the 'vioience of the gaze' in order to witness women's violence and its elision in a way that pays attention to theory but doesn't simply reduce violence against women *to* theory, that pays attention to history but doesn't simply trope violence against women as a metaphor *for* history? This book responds to these questions by attending closely to the local histories of the violent acts it investigates as well as to those acts' implications for contemporary performance theory and practice. I juggle both trajectories but work hard never to collapse them, never simply to 'claim' the early moderns as our contemporaries or the rape or abuse experiences of early modern women as our own. I aim to historicize violence against women as a process of performative disappearance at one particular moment in cultural history – a moment to which we return again and again, obsessively, in our own theatres, as though desperate to see again what we haven't quite seen before. In the process I aim to interrogate not only early modern English culture's effacement of violence against women but also our own encounter with that evasive history: our resistances to it, our attempts to witness it, and the meanings of those attempts for the way we go to the theatre and do performance scholarship right now.

The violated female body is the blind spot at the heart of Phelan's *Unmarked*; it is the one miss the book cannot assimilate into its critical framework. The scar Festa's body leaves in Phelan's text suggests that the woman in pain is the litmus test for the unmarked as a theory of absence-as-representation, and for the theory of performance-as-witness it inaugurated. To reckon with the in/visible act of violence against women is, perhaps, to realize the unmarked's full potential as both performance theory and performance ethics.

In/visible acts

After this very theoretical introduction my remaining chapters look closely at a number of early modern cultural and dramatic texts alongside contemporary performances of those dramatic texts in order to elaborate the in/visible act in practice. Chapter 2 explores the performative politics of early modern instruction for rape victims, its brutal rending in Shakespeare's *Titus Andronicus*, and its contemporary uptake in Deborah Warner's 1987 production of the play for the Royal Shakespeare Company and in Julie Taymor's 1999 film, *Titus*. I argue that the 'hue and cry' that follows rape is designed to stage a cultural rehearsal of the act in the public street, one that has the effect of divorcing rape from the body of its victim. Lavinia, left without tongue or hands by her violators, is unable to complete the performative transfer the hue and cry demands; as a result she can only perform its radical failure, an unexpected in/visible act that has the disorienting effect of generating something of what that transfer can never fully manifest in representational space. Chapter 3 begins with a woman who prophecies her death in childbirth and then, falling ill, appears to orchestrate the deathbed performance of her own salvation, deflecting attention from the material conditions of her labour toward heaven's grace and God's purpose. The chapter goes on to explore the ways in which early modern household and marriage manuals written by religious officials encouraged wives abused by their husbands to stage just such a performance of salvation in order to recuperate the culturally disruptive image of the husband who takes the tenets of 'reasonable correction' too far. Reading Thomas Heywood's *A Woman Killed With Kindness* alongside one of the most popular (and notorious) marriage manuals of its day, Whately's *A Bride-Bush, Or, A Direction for Married Persons*, I parse the play's 'kindness' for its response to the social literature's own elision of 'kindness' and 'kill'. I then explore Katie Mitchell's 1992 Royal Shakespeare Company production, arguing that her practice of radical naturalism allowed her actors to shock Anne Frankford's (supposedly) seamless performance of salvation late in the play with its own hysterical disavowals, manifesting the in/visible act latent in Heywood's final scene. Chapter 4 returns to *The Duchess of Malfi*. I map the play's resistive engagement with Protestant martyrology as Webster stages the Duchess's critical despair in the face of her torture and her rage against her posthumous re-scripting as a self-effacing female martyr. My contemporary intertexts here are Peter Hinton's 2006 Stratford Festival production and Phyllida Lloyd's 2003 Royal National Theatre staging. Reviving the critical potential of the Duchess's eerily sentimental return as 'echo' in Act Five, I parse the role ghosting plays in both productions, engaging audiences as a visual, indeed visceral part of the show in order to shape our responses to the play's trenchant spectacles of failed witness. I end the book with a reading of *The Changeling*, a play in which narrative colludes with architecture as a woman's violation is made to disappear before our very eyes. Beatrice Joanna is a classic Jacobean 'bad girl', but she is also a rape victim – a paradox virtually incomprehensible to her cultural moment. The non-spaces on which this play's debt to sexual violence is structured – the interval between acts in which De Flores rapes his prize; the 'prospect from the garden' by which Jasperino convinces Alsemero of Beatrice Joanna's guilt; Alsemero's closet, in which Beatrice Joanna is forced to endure further rape as well as brutal murder – enable the ongoing percepticide (the rape's insistent forgetting) that guarantees *The Changeling*'s modern legacy. This chapter ends the book with an extended reading of how the spaces of performance – both within *The Changeling* and in modern production – collude with spectatorial desire to produce a very specific, very retrograde view of Beatrice Joanna as villain and vixen rather than victim. Cheek by Jowl's touring production of the play, set in a madhouse-cum-rehearsal hall and staged literally backstage at London's Barbican Centre in the summer of 2006, allows me to examine this terrifying text's hidden potential to produce what I call an architecture of feminist performance.

A word about methodology. Although this book is intimately bound up with the pleasures and pressures of representational history, I do not engage in extended close readings of historical texts. I gather and synthesize the superb work of many feminist historians of early modern culture in order to elaborate the social, political, and religious conditions that inform the subjects this book handles; when my analysis warrants, I also explore specific historical texts in relation to the play texts and performances I read in depth. This hybrid approach allows me to use the historical scholarship already in circulation to advance new readings of select historical trends as acts of cultural performance, and to compare those acts with their reflections on the early modern stage. Finally, because this is not just a book about early modern culture but also a book about contemporary acts of theatrical engagement, I implicitly imagine an audience for the productions I explore. This audience is imperfect and subjective (I insert my own watching self whenever possible, conjuring my memory of the viewing experience), but it is also theoretical and ideal; like Jill Dolan, David Román, Roger Simon, and Alice Rayner, I am writing towards hope, towards the potential performance embeds to stage an encounter with the past that opens up a new view of the present and the future. Nevertheless, I recognize the risks of such an approach; as Roberta Barker writes, 'The politically engaged critic cannot indicate what whole audiences thought about a production; she can only trace her own interventions with those components of a performance and its reception that her temporal and ideological positions allow her to see' (2007, p. 24). While I must rely on my own skills as what Barker calls a 'politically engaged' spectator to flush out moments in text and in performance that react to early modern conceptions of violence against women and gesture toward the in/visible act, I also hope that my readings will be scrutinized and tested against other readers' and spectators' responses. To see the ghosts of representation is really to see with one's imagination; that imagination leaves room for numerous other acts of politically engaged play. Such is the pleasure of scholarship; such is the pleasure of theatre.

This book seeks a contemporary afterlife for early modern violence against women. It deals in canonical dramatic texts from the period that are still regularly performed today, and it understands the resulting performances as complex cultural documents that do not just rehearse long-past, irrelevant social behaviours from another world, but actively negotiate the import of those behaviours as part of an ongoing feminist history, and theory, of the representation of violence against women. For anyone who has ever said Shakespeare and his fellow Renaissance men were our 'contemporary': I ask what it means to claim their work as 'ours', to take on their history as our history, and to witness the acts of violence against women their plays (don't quite) stage, right now.

2
Rape's Metatheatrical Return: Rehearsing Sexual Violence in *Titus Andronicus*

In the May 2004 issue of *Theatre Journal*, Jody Enders tells the story of one Mrs Coton who, on the night of 1 May 1395, was gang-raped in Chelles, France, by a group of men in town to see a show. The story is not new; in fact, it is 'one of the foundational narratives of medieval French theatre history' (p. 181). What is new is Enders's emphasis on Mrs Coton as the victim of an act of violence tied, troublingly, to an act of representation. Enders goes on to argue that rape's normative, even morally instructive stage representations in the medieval period may have proved an important standard by which the crime was 'habituated' in broader legal and social arenas, and vice versa. Then, in an important aside to contemporary performance scholars, she asks: 'what are the moral ramifications of theatre's own scripting and virtuality?' (ibid., pp. 179–80; 167).

I rehearse Enders's thinking as prelude here because its own ramifications are much further reaching than its narrow historical context may at first imply. Enders is not just reclaiming the rape of Mrs Coton from scholarly erasure; she is asking theatre historians, theorists, practitioners, and audiences to pause for a moment over the ethics of the performance event and to ask ourselves for what, exactly, theatre is willing to take responsibility. At what point do our basic assumptions about the essential difference between rape as 'real' and rape as 'only' representation begin to break down? When and how does seemingly innocent watching tip over into a kind of doing, into what Taylor calls 'percepticide'? How do we account for those moments when theatre poses genuine risk to certain publics – both in the spectacle it frames and in the forms of spectatorship for which it tacitly calls? And, finally, how do we build a response to that risk into our own work as scholars and practitioners?

This chapter explores the theatricalization of rape during the late Elizabethan and Jacobean periods, a world two centuries away from that fateful night when Mrs Coton makes theatre history. Like Enders, I am interested in the moments when theatre and sexual violence unexpectedly converge: I argue that early modern culture constructs rape victims as actors in a theatre of trauma built to externalize and thereby instantiate the crimes they are ostensibly reporting. I call this phenomenon rape's metatheatrical return to denote its rehearsed quality, its necessary and often palpable scriptedness, and the essential familiarity of its theatrical framework for both its actors and its audiences. While for each rape victim called upon to rehearse the script of rape's telling the performance is a first, its efficacy derives from the fact that, for her as for her witnesses, it is not meant to be new at all. We've seen this performance many times before; it derives its power from our fluency with its terms. It functions only insofar as it references the *last* such performance we saw, the last time a victim successfully made her rape (re)appear. I similarly use the term 'metatheatrical' to describe the anticipated social efficacy of rape's public return. Like Hamlet's *Mousetrap*, the rehearsal of rape is understood to be a social good, a means to call out a crime. But as trauma past becomes performing present, what consequences arise? Rape's metatheatrical return is an event constituted after the (f)act in carefully arranged word and gesture, to be performed for the edification of an authorizing public *before* any legal appeal can be made; that is to say, it is an exposure that is simultaneously an elision. It is designed expressly to translate rape's meaning, its suffering, and its cultural import from the space of a traumatized woman's body into the space of the male public sphere. As it turns rape into a tangible, actionable thing, the metatheatrical return raises difficult questions about how, and if, rape can be seen and known, and about how we might reframe the return in contemporary performance in order to mount a more ethical engagement with sexual violence in the early modern texts we choose to study and to stage.

Reverberations between the performance of rape on and off the early modern stage suggest that the law and the theatre may well have been mutually informing where sexual violence was concerned. Rape is never represented directly in early modern drama (Catty, 1999, p. 138), and its banishment from the boards necessitates its carefully structured reappearance after the fact. The wronged maiden typically takes her cues from a clutch of classical heroines, appearing to her witnesses as a contemporary Lucrece, Virginia, or Philomela. She enters 'ravished' (Shakespeare's Lavinia[1]) or 'unready' (Heywood's Lucrece [l. 2090];

both directions imply dishevelled garments and appearance), laments her shame, calls for death, or reaches for the knife herself. I refer here specifically to rapes in which a chaste, idealized woman is raped or threatened with rape and embraces death in order that she may be absolved and the plot may be brought to an appropriately heroic conclusion; Shakespeare's Lucrece (1594) – who speaks at eloquent length of her sorrow, shame, and anger before committing suicide before her male witnesses – may be, despite her roots in poetry, the icon to whom many a ravished maiden aspires on the early modern stage.[2] Female characters whose own sexual desires complicate an audience's reading of their assaults (Bianca in *Women Beware Women*; Beatrice Joanna in *The Changeling*), or female characters who react to their rapes in rage rather than in guilt, refusing to take the blame for the crime into their own bodies (the daughters in *Bonduca*), tend not to conform to this narrative structure and are not meant to elicit our sympathy.

The stage heroine's use of standard symbolism and rhetoric and her plain citation of chaste mythical precedents serve to locate her within an ongoing extratheatrical narrative about what rape means and how it should be reported: all her public words and gestures work to confirm her innocence and remind her male friends and relations of what they must do next to avenge her/their honour. Indeed, the most likely reason for rape's stage absence in the early modern period is Renaissance culture's primary focus not on the pain of violation, but on the difficulties of action, honour, justice, and revenge; rape's dramatic representation among the early moderns reflects not what a modern audience might understand about the experience – the victim's heinous bodily and psychic suffering – but rather what rape means to those to whom it is reported, who can access it only as vicarious witnesses but who also bear the heavy responsibility to absolve the victim of any potential complicity and to mobilize the force of the law (see also Catty, 1999, pp. 22, 108–9). Rape is staged in early modern England as it is made known in the world beyond: as a rehearsal for public confirmation, one that has the scent of the playhouse always about it.

The imbrication of performance in confessing and proving rape in the early modern period may appear, at first glance, to suggest the opposite of Enders's argument. Theatre seems far more hero than villain in this picture: the assaulted woman shows and tells her wrong and the men for whom she performs put their faith in her theatrical proof and lead her to justice. The risk of theatre for the early modern rape victim, though, is finally no less than that suffered by Mrs Coton: as rape becomes theatre, its consequences for her body, her psychic well-being, even her future prospects fall out of the frame. The question that worries me: can we, should we, recuperate this loss? If so, how? In the first part of this chapter I explore the legal advice offered to early modern rape victims and their families as they sought to report sexual assault and make an appeal, and I wonder: how does theatre work here, and how might it work differently? In the chapter's second half I venture one possible answer to that question by turning back to the theatre, to the infamous violation of Lavinia in Shakespeare's *Titus Andronicus* and its uncanny troubling of the conventions of rape's metatheatrical return. Lavinia, stripped of hands and tongue, becomes an accidental guerrilla performer in Shakespeare's traumatized Act Three; she offers her family not the story of rape they need to hear in order to seek revenge, but the story the return is designed to mask, its corporeal disappeared. I then invoke the troubling of *Titus Andronicus* by two contemporary women theatre artists: Deborah Warner and Julie Taymor. Warner's 1987 production of the play for the Royal Shakespeare Company and Taymor's 1999 film adaptation each take very different approaches to staging Lavinia's in/visible act. Warner produces rape's feminist realism, while Taymor shies away from any specific ideological influences or overarching theatrical tropes in a movie that has been lauded as clever pastiche but also condemned as an apologist for violence. Although Warner might appear to offer the model feminist performance of sexual violence, I close the chapter by advocating for Taymor's interpretation as a more provocative response to the history of rape's theatricalization. Unlike Warner, who focuses incessantly on Lavinia's bodily response to trauma, Taymor uses the play's complex metatheatricality to bind Lavinia's experience of violence to the expectations of performance, exposing the sinister processes of the return even as she celebrates the enduring plasticity, the malleable potential, of the stage.

The politics of 'hue and cry'

Rape creates a fundamental epistemological dilemma. Victims 'shut down' (Konradi, 2007, p. 21), fold inward, enclose their stories of suffering within their flesh, bones, hearts, and brains. Mieke Bal writes: 'rape cannot be visualized because the experience is, physically as well as psychologically, *inner*. [...] In this sense, rape is by definition imagined' (2001, p. 100). Rape, Bal argues, can only appear in social space as an aftershock, 'can only exist [...] as *image* translated into signs' (ibid., p. 100). It can only be made real, be made to matter (made a legal or a social matter) at a distance from the suffering body, as a carefully codified representation of an arguably inaccessible event.

For raped women, however, the experience is not imagined but unimaginable. It is a shattering of the self, to borrow Cynthia Marshall's provocative phrase, the unravelling of one's ability to be safe, to find a home, in the world (Konradi, 2007, pp. 34–6). It is perennial, disorienting fracture, constant loss. In her stunning 2002 book, *Aftermath*, survivor and Dartmouth philosophy professor Susan J. Brison remembers the moment when that sense of fracture hit home where she both works and lives:

> Ten years ago, a few months after I had survived a nearly fatal sexual assault and attempted murder in the south of France, I sat down at my computer to write about it for the first time and all I could come up with was a list of paradoxes. Things had stopped making sense. I thought it was quite possible that I was brain-damaged as a result of the head injuries I had sustained. Or perhaps the heightened lucidity I had experienced during the assault remained, giving me a clearer, although profoundly disorienting, picture of the world. I turned to philosophy for meaning and consolation and could find neither. Had my reasoning broken down? Or was it the breakdown of reason?
>
> (p. ix)

Brison mines the paradoxes that emerged from her experience of assault and uses them to build an ethics of response – of witness – to rape. *Aftermath* pays heed to the ways in which rape trauma 'haunts' the mind, the body, and the senses of victims, ready to return anyplace, anytime (ibid., p. x; see also pp. 15 and 20). But it also insists on the absolute necessity of talking about that trauma, 'again and again' (p. 16), of taking hold of the reigns of representation. Brison is aware of the challenges embedded in the attempt to 'speak about the unspeakable without attempting to render it intelligible and sayable' (ibid., p. xi); her book is a testament to the importance of 'finding a language that is true' (p. xi) to the profound, world-breaking self-loss that marks the experience of rape, the pain of which is increased significantly by the extraordinary cultural amnesia it inspires. Brison writes extensively about the personal, social, and political consequences of rape's forgetting in contemporary Western culture. Remembering her own family's inability to communicate with her in the aftermath of assault, she argues: 'In the case of rape, the intersection of multiple taboos – against talking openly about trauma, about violence, about sex – causes conversational gridlock, paralyzing the would-be supporter. We lack the vocabulary for expressing appropriate concern, and we have no social conventions to ease the awkwardness' (ibid., p. 12). Rape *is* unimaginable, but for so many reasons – justice; healing; a return to community; the advancement of women's human rights – it must be imagined, must be spoken, must be signed. The question is how.

In early modern England, the anxiety circulating around the defining lack of an objective proof of rape, combined with misogynist prejudices against women's sexual independence, meant that the legal burden fell directly upon victims to prove rape's occurrence by first proving their own sexual innocence. By the time Elizabeth's reign neared its end, rape was defined by the absence of a woman's consent to sexual intercourse with any man who was not her husband. A substantial amount of feminist legal and literary scholarship has already dealt with rape's notoriously mutable historical definitions, its persistent tendency to be classed as a property crime, and the evolution of a woman's refusal of consent as essential to fixing rape among the early moderns. A brief rehearsal of this scholarship's principal findings will allow me to set the stage for my own reading, below, of early modern instructions to rape victims in light of those instructions' compelling theatrical resonances.

For a significant portion of the thirteenth through the early sixteenth centuries rape was largely a property crime, defined by chastity's theft from a woman's family with or without her consent (Burks, 1995, p. 765; see also Post, 1978; and Phillips, 2000). The early thirteenth-century Bracton treatise equates rape with defloration for the first time (Bracton, 1968, pp. 414–15; Phillips, 2000, p. 132); the first statute of Westminster (*c*.1275) similarly elides rape with the abduction of a marriageable woman by, or her or elopement with, 'an unacceptable suitor' (Post, 1978, p. 158). In her recent reassessment of medieval rape law, Kim Phillips helpfully notes that the Glanvill treatise of 1187 had earlier placed particular emphasis on the raped body as bleeding or assaulted, and thus on rape as, if not a specifically sexual crime, without question a corporal one (2000, p. 129). As time passed and the interests of the gentry gained legal and political prominence, legislative emphasis began to shift away from the physical experience of the rape victim and onto her symbolic commodity value (ibid., p. 141). By 1285 rape was fully enshrined as a property crime, and remained so until statute changes immediately before and near the end of Elizabeth's reign, in 1555 and 1597, began once more to focus rape's specificity on the body – but this time on the sexualized body, and specifically on the absence of a woman's sexual consent in determining the crime. Nazife Bashar has argued that this new focus helpfully reframed rape as 'a crime against the person' (1983, p. 41), but in practice the old emphasis on property remained, to be awkwardly negotiated alongside the new emphasis on women's agency: Bashar notes that convictions remained most likely in cases where rape's economic consequences were most clearly tangible (ibid., p. 42). In fact, the sixteenth-century emphasis on a woman's non-consent as the ultimate arbiter of rape, combined with misogynist assumptions about women's unruly sexual appetites, resulted in renewed fears of what Deborah Burks calls women's sexual 'defection'. Was she raped, or did she give herself (her husband's/father's property) away? Failing to prove non-consent, a woman became not only complicit in the crime but, in popular prejudice, her own rapist (Burks, 1995, pp. 766, 770).

So rape comes, by the turn of the seventeenth century, to rest on the proof of female sexual innocence, a remarkably challenging prospect given that a rape victim not only had to refuse consent to her attacker but also had to find a way to make her non-consent visible to a public deeply sceptical of women's sexual motives. In *Common Bodies* (2003), one of the most comprehensive discussions to date of women's experiences of consent and desire in the early modern period, Laura Gowing argues that the more sexually legitimate a woman was perceived to be, the less she was able to speak of sexual assault directly (see also Walker, 1998, pp. 5, 8). A married woman reporting rape would be expected to displace her experience of sexual violence onto household goods and chores, explaining the assault in domestic metaphors (ibid., p. 93; see also Chaytor, 1995, pp. 384–5).[3] Gowing also notes, however, that *outside* of the framework of formal rape trials at the assizes, where rape was a capital offence, women were granted far greater licence to speak about what plainly amounted to sexual assault. In the quarter sessions or church courts, a woman accused of fornication or illegitimate pregnancy might describe a 'seduction' that was obviously violent; because, however, the official matter at hand was not rape, the explicitly violent nature of the reportage was merely taken for granted as a quality of sex. Women's sexual passivity was assumed at this time, just as men's sexual appetites were imagined to be inherently aggressive; as a result, the elision of sex and violence in both legal and popular discourses was routine. Gowing concludes: 'As a crime, rape was difficult to prove, unlikely to be pursued, and hard to testify about. But the violations of sexual assault and forced sex were familiar social facts as well as personal experiences' (2003, p. 101).

Gowing demonstrates the extent to which rape's telling was both absolutely expected and yet practically impossible in early modern England. To advocate effectively for herself, a victim needed a very clear understanding of the positioning of both women's bodies and women's voices in the period's social spaces. She also had to adhere to a clear script, with clear precedents. Telling rape in early modern England was a conscious gender performance, a careful citation of standard tales of assault already proven both believable (in the public square) and actionable (in the courts of law). *The Lawes Resolutions of Womens Rights*, a 1632 legal tract written with women's legal needs specifically in mind, reveals that a victim's successful appeal to the authorities relied on two separate but related actions: verbal complaint combined with physical demonstration.

> [S]he ought to goe straight way, [...] and with Hue and Cry *complaine* to the good men of the next towne, *shewing* her wrong, *her garments torne*, [...] and then she ought to goe to the chiefe Constable, to the Coroner, and to the Viscount, and at the next Countie to enter her Appeale, and have it enrolled in the Coroners roll: and then day was to bee given her, till the comming of the kings Justices, before whom she was againe to re-intreat her Appeale [...].
> (*Lawes Resolutions* 5:30, pp. 392–3, my emphasis)

Lawes Resolutions prefaces the above instruction by citing the influential Bracton treatise (*c.*1218–29), but the scene has several legal precedents including Glanvill. At its base, its advice is stock: by 1632 victims of rape had been told for centuries to raise the hue and cry by showing their physical injuries to 'the good men of the next towne' and the authorities beyond. Glanvill writes:

> In the crime of rape a woman charges a man with violating her by force in the peace of the lord king. A woman who suffers in this way must go, soon after the deed is done, to the nearest vill and there show to trustworthy men the injury done to her, and any effusion of blood there may be and any tearing of her clothes. She should then do the same to the reeve of the hundred. Afterwards she should proclaim it publicly in the next county court; and when she has made her complaint, the form of proceeding to judgment shall be as stated above.
> (1993, 14:6, pp. 175–6)

Bracton's wording differs only slightly,[4] but also expands significantly on Glanvill by including the specific language a woman should use for her appeal (which *Lawes Resolutions* cites in 5:32 and 5:34, pp. 394–6) and by emphasizing the authority of the woman's voice in the process of that appeal:

> She must go at once and while the deed is newly done, with the hue and cry, to the neighbouring townships and there show the injury done to her to men of good repute, the blood and her clothing stained with blood, and her torn garments. And in the same way she ought to go to the reeve of the hundred, the king's serjeant, the coroners and the sheriff. And let her make her appeal at the first county court, unless she can at once make her complaint directly to the lord king or his justices [...]. Let her appeal be enrolled in the coroners' rolls, *every word of the appeal, exactly as she makes it,* and the year and day on which she makes it. [...] The words of the appeal are these: 'A., such a woman, appeals B. for that whereas she was at such a place on such a day in such a year etc. (as above) (or "when she was going from such a place to such," or "at such a place, doing such a thing") the said B. came with his force and wickedly and against the king's peace lay with her and took from her her maidenhood' [...].
> (1993, pp. 415–16)

Both Glanvill and Bracton expect a standard verbal proclamation of rape at the formal appeal stage, but they also place primary emphasis during rape's initial revelation on a victim's 'show' of – specifically *physical*, plainly *visible* – injury to a very specific category of witnesses: reputable men. Both treatises thus suggest that visual evidence carries the burden of proof in the telling of rape, and that subsequent official proclamations are only made possible by that initial visual evidence and, crucially, by its ability to compel the belief of the 'good men' of the town. Glanvill makes no mention of a woman's speech before the official appeal stage ('A woman who suffers in this way must go, soon after the deed is done, to the nearest vill and there *show* to trustworthy men the injury done to her'). Bracton, as I note above, places particular emphasis on the very specific language of a woman's eventual appeal, but only after prescribing her apparently silent show of blood, stained clothing, and torn garments. Performing the hue and cry first image- and *then* letter-perfectly was a serious, even life-or-death matter for injured women and their families: Barbara Hanawalt's research suggests that a victim's failure to conform to the scripts set out in treatises like Glanvill or Bracton could have resulted not only in her appeal's failure, but also in her subsequent prosecution for making a false appeal (1998, p. 127). No wonder, then, that these texts prescribe a burden of visual proof, followed by a carefully worded, time-worn set of codes for speaking of the crime: both allow women to make their fundamentally 'inner' (Bal, 2001) experience of violation conform exactly to (male) cultural expectations of what rape 'should' look like on the outside.

On the surface, *The Lawes Resolutions of Womens Rights* appears to be not much more than an updated, modernized (the text itself is written largely in English) version of Glanvill and Bracton, and yet the combined emphasis on show *and tell* in the initial stages of rape's appeal in *Lawes Resolutions* makes for a subtle but significant shift: it encourages women to body forth the implicit theatricality of – the *performance of innocence* implied by – the scripts in the earlier treatises. I read this shift as essential in order for the text to frame the hue and cry for an early seventeenth-century readership. Such a readership would have been familiar with the stock advice of older legal tracts, but also with the conventions surrounding the representation of sexual violence in the newly emergent private and public theatres. In the wake of Calvinist influence over the English Reformation – which heralded religious and secular worries over the transparency of image; misogynist fears of women's speech; fears of women's decorated bodies as sacrilegious icons (Zimmerman, 2005, pp. 55–6); and a renewed emphasis on text, writing, and reading as an arbiter of public and private truth (Coats, 1992, pp. 3–4) – the theatre became one of the chief sites at which new epistemologies were forged in the early modern period. Against the Reformation's anxieties over the impoverishment of image on one hand and its often totalizing claims about the power of the Word on the other, drama argues that spectacle – the marriage of spoken word and image – is a legitimate 'way of knowing' (O'Connell, 2000, p. 144), a pragmatic alternative to a short-sighted reliance on any one sign system. Theatre might not always show you the truth, but it can (as Hamlet's *Mousetrap* memorably suggests) use the intersection of visual and aural claims to instantiate (or interrogate) the legitimacy of each.

Given this epistemic framework, the slightly altered structure of the hue and cry in *Lawes Resolutions* makes pragmatic sense. Making rape known, it argues, is a matter of staging an increasingly familiar process of show and tell that relies on the mutual reinforcement of word and image people will recognize from, among other places, the theatre. In *Lawes Resolutions* rape's revelation becomes more overtly a staged tragedy, the victim a tormented heroine: she is to go to the nearest town and raise the hue and cry by 'complain[ing]' of the crime while *at the same time* 'shewing her wrong', her ripped clothes, and bodily wounds in the public street. The *Oxford English Dictionary* defines 'complain' at this time as a synonym of 'bewail' and 'lament', especially in pain or suffering. (The term's earliest such use is given as 1374, approximately 150 years after Bracton and 100 years after Westminster I.) It may also refer to breast-beating, which, significantly, is how Titus Andronicus encourages his daughter Lavinia to express her sorrow after she has been robbed of the ability to speak of her rape directly. *Lawes Resolutions* thus calls for a particular kind of dramatic, even tragic speech as 'hue and cry': a victim's damaged body, spoken *through* her lamenting voice, will sign the truth of her chaste narration, producing her innocence via the theatrical convergence of action, speech, and sight.

The *Lawes Resolutions*, like Glanvill and Bracton before it, describes (prescribes) rape's theatricalization as a process of making rape known, but it is actually a process of making rape *real*, a substantiation as well as a translation. Only after a victim has properly performed her trauma for 'the good men of the next towne' can she sue to the authorities; the performance *makes the crime actionable*. Just as in the theatre, where the heroine's 'return' to the stage post-rape stands in for an act that exists only as a gap in the story, the carefully scripted telling of rape transforms the crime from an event with little epistemic status (an event *of* the woman's body) into an event *of* public space, and *for* the men who guard that space. Rape's metatheatrical return is not just a re-enactment of an event that has passed; it is a 'rehearsal' that, for public purposes, is an original, an event that can be witnessed and therefore subsequently may be prosecuted. If rape is fundamentally unmarked, a crime that trumps the (eye) witness, then it is ultimately much more than an assault on a body or family. It is an assault on knowledge itself. By reproducing her rape as a performance in which a standardized version of her suffering registers as proof of chaste intention before a body of citizens designated as her official witnesses, the victim mitigates the anxiety born both of those witnesses' failure to see her original trauma, and, beyond that, of the impossibility of their ever really knowing the truth or falsehood of her claim of non-consent. And here, in turn, lies the paradox of the metatheatrical return: rape is (re)staged in order to allay fears about its initial, invisible enactment, but in the process it goes missing again, metamorphosing from a woman's psychic and social trauma into a matter of masculine honour, an occasion for heroic deeds undertaken by wronged fathers, husbands, and brothers (their interests supported but not displaced by the victim's essential testimony) in the defence of family names and family purses. Theatre is a double-edged sword: it both resolves the problem of rape – it cannot be prosecuted if it is not known, and it cannot be known if the victim refuses to raise the hue and cry – and effaces it at the same time, solves it, in fact, *by effacing it*.[5] Made into spectacle, rape slips away from the woman it haunts as its invisible effects appear to disappear.

As I have developed and presented this argument before various audiences over the last few years, I have been reminded again and again that not much has changed. The politics of hue and cry are as fraught as ever in my own world: hospital rape kits substitute for the ritual display of ripped dresses, while updated cultural scripts replace Bracton's carefully worded standard appeal (Rayburn, 2006). Police, judges and juries continue to disbelieve a substantial portion of appellants: even in apparently 'straightforward' cases officials often look for ways in which to recognize and represent victims as officially non-consenting (Jordan, 2004, pp. 177–214; Brison, 2002, pp. 7–8), while in all cases rape trials hinge less on the letter of the law than on what Corey Rayburn (2006) presciently calls the 'burden of performance' – 'the narrative structure most appealing to the jury'. Rayburn writes:

> When complainants testify, they assume roles that put their gender identity into question. [...] When a complainant is telling her story, she must impress a jury that has been inculcated with a lifetime of rape imagery and accounts, making the burden of performance a substantial impediment. Accusers must convince a jury, jaded by rape stories and pictures, that her story is 'special' enough to warrant a guilty verdict. The end result is that prosecutions are doomed to fail in most situations.

Rayburn's words, woven through my analysis of rape scripts above, suggest that, in the pre-and early modern periods as today, the public rehearsal of rape relies for its functionality upon a well-wrought reality effect. Not only does the victim need to fill the gaps left by the originary act of violence, but she must also do so without calling undue attention to the fact that she is performing a preordained role, playing both herself and the self she is expected, hoped, but feared not to be. Although the conventions of the early modern metatheatrical return were so well known as to be self-conscious, the performance itself had to be seamless. Young women would have to be 'coach[ed]', lest they erred in their testimony or failed to render it believably (Hanawalt, 1998, p. 128). For the 'good men' of the nearest town as for the good men and women of the modern jury, the return mimes a web of standardized beliefs about how women are in the world for men, staging an ideal in order to ward off the possibility that they might not be in the world for (their) men alone. Rape's metatheatrical return is thus a perfect example of what Elin Diamond has called 'patriarchal mimesis' (Diamond, 1989, p. 64), the process by which woman serves as man's mirror-up-to-culture, reflecting his image of his own centrality by masking his pernicious absence from – and her complex, multiply-vectored presence at – the scene of representation.

Patriarchal mimesis, a product of Platonic prejudices and Aristotelian teleology, demands that mimetic copies correspond exactly to their models, pointing to the irrevocability of the model's cultural primacy. Following Luce Irigaray's playful excavation of Plato's cave in *Speculum of the Other Woman*, Diamond doesn't accept the model-copy correlation, preferring instead to see the resulting mimetic structure as inherently unstable. Mimesis is always a residual act: it leaves a trail of 'performance remains' (Schneider, 2001) pointing to the choices it has made in constructing its universe. Likewise, the theatre is a fundamentally residual space: grounded in an economy of substitutions, theatrical events always contain both more and less than we ultimately hear and see. So what happens, Diamond asks, when the system breaks down? What happens, we might ask in turn, when the symbols, the stock narratives, the tropes, and echoes by which the early moderns understood rape begin to grow excessive, overtly self-referential, or appear as somehow oblique, alienated from their anticipated referents? Obviously, the question is far from theoretical for a rape victim and her family, since failing to realize the return would have had serious consequences for early modern women. But, at the same time, the theory probes a possibility: what if, somehow, the underlying logic of the return crept back into its performance? What might it reveal to a sympathetic audience? Could it allow us to craft an ethical, contemporary response to the early modern history of rape's representation, and to make space for audiences to witness not just that history's lost bodies, but theatre's own culpability in their disappearance?

The fragmentation, the disorientation, and the gaping sense of loss that attend the experience of rape in the moment of its happening and in its immediate aftermath are exactly what rape's rehearsal must evade even as it brings rape, in surrogacy, back to life. I do not want to imply that there is no room for the articulation of any 'deeper' feeling within the structure of the metatheatrical return; surely many early modern women found a way to convey something of their bodily and psychic horror in the process of saying and doing all the right things for the authorities. Nevertheless, the script offers little room for improvisation. The sanitized symbols of loss – blood, torn clothes, pre-planned narrative – organize the terrorizing confusion that attends sexual violence into a coherent metonymy that can sign outward from the body; this is the groundwork of the return's patriarchal mimesis. In order to re-frame it as a potentially ethical performance of sexual violence, a feminist performance of sexual violence, we need to ask: what would it mean to stage the metatheatrical return *as fracture*, as the residue of violence *unassimilated* into convention? Although, as I noted in my introduction, classical feminist performance theory investigates the violence of the gaze, the metatheatrical return invokes less a totalizing gaze than a collective blink, a strange refusal to look embedded within a pernicious, obsessive need to see. Rape's rehearsal lives on the edge of invisibility; a feminist strategy for politicizing it must play upon that edge, play with the 'unmarked' experience of sexual violence that goes missing within the tightly circumscribed terms of the return's performative transfer. What would it mean to stage rape as missing, as loss – the loss occasioned by sexual trauma as well as the loss of specular power? What would it mean to stage rape as the impoverishment of the witness the crime threatens and the return covers up? How might we recaste the victim/performer neither as pure body (the feared) nor as pure symbol, idealized icon (the domesticated), but as 'the emblem of a body' gesturing uncannily, desperately, angrily at the limits of the known, the knowable (Phelan, 1997, p. 82)?

The public square is too high-stakes a place for such an experiment, such an unexpected, inexplicable hue and cry. To find performers willing to take the risk we need to turn back to the theatre proper, to the secular, self-conscious, early modern stage.

Messing up (in) the theatre of sexual trauma

Titus Andronicus is famous for violence, but Lavinia is not simply famous for being raped. She is famous for out-Philomela-ing Philomela. Appearing 'ravished' on cue at the end of Act Two, Lavinia finds herself unable to enact the metatheatrical return: having lost both her tongue and her hands to her rapists, she lacks the means either to show or to tell. She is a spectacular puzzle, a cipher of loss pointing incessantly, in her terrified desperation to find a way back to language, to the gap in her mouth and the stumps at her wrists. A body riddled with holes, a body that *becomes* a hole, Lavinia reveals the macabre futility of forcing a rape victim to perform on spec in order to impose public coherence upon a fundamentally disordered, deeply disorienting bodily and psychic experience. *Titus Andronicus* came of age as a text of literary worth with the rise of poststructuralist theory. The play's splendid sensationalism has always been a favourite of the stage, from its first London performances in the late sixteenth century through a number of adaptations in several languages, including Edward Ravenscroft's for the Restoration stage in 1687. Even the Victorians found a way to play the play, though they needed to excise the rape and mutilation of Lavinia completely in order to do it. Literary critics, however, have been far less kind to *Titus*; until the second half of the twentieth century it was deemed good only for gore (Bate, 1995, pp. 33–4, 48–59). *Titus*'s stock rose after Peter Brook's critically acclaimed 1955 production starring Laurence Olivier and Vivian Leigh, but the late-century explosion of scholarly interest in the play came to rest on its penchant for punning and wordplay, its obsessive meta-textuality prismed through its obsession with the fragmented body.[6] Lavinia, not surprisingly, is the silent centre of this critical action, and the substantial body of feminist criticism on the play has staged a protracted debate about whether or not she may be called an author when she finally scratches the names of her rapists in the sand at the beginning of Act Four.[7] This feminist work has ultimately been essential for Shakespeare criticism: it not only improves *Titus*'s canonical standing but also complicates the standard portrait of Lavinia as silent sacrifice as it thinks through the problems of agency embedded in the character. Nevertheless, in their relentless focus on the image of the (deconstructionist) text in *Titus Andronicus*, scholars raised on Barthes, Foucault, and Derrida have obscured the play's equally compelling focus on performance. In its latter half *Titus Andronicus* is self-consciously histrionic, mixing madness with acting in an often-uncertain mélange that prefigures the more carefully wrought machinations of that most metatheatrical of Shakespeare's revenge tragedies, *Hamlet*. And, just as she is the centre of *Titus* criticism's textual obsessions, the touchstone of all this playing on play is Lavinia herself.

What gets lost when critical eyes turn away from *Titus Andronicus* as a product of and for the theatre, of and for a culture that increasingly understood itself in terms of performative acts undertaken in the public realm? Pascale Aebischer argues that 'the literary aspect' of Lavinia's rape, its longstanding 'association with intertextuality and rhetorical tropes' for directors and scholars alike, 'constitutes the greatest challenge to an "embodied" consideration of [that] rape as an act of physical cruelty perpetrated by men on the body of a woman' (2004, p. 27). For Aebischer, the relentless focus (both in the play itself and in the critical literature) on 'reading' Lavinia makes it almost impossible to *see* Lavinia as the battered and destroyed woman that emerges to receive Marcus's long blazon when he discovers her damaged body at the end of Act Two, and harder yet to recognize her as a performer (rather than, say, a silent metaphor) in the dysfunctional family drama that takes over the stage during Act Three. Ironically, given Lavinia's lack of hands and tongue, the textual trumps the visual in her aborted hue and cry: faced with a niece both perversely unrecognizable and viscerally troubling, Marcus points her body toward literature, calling her Philomel as he attempts to fill the uncanny gap her bleeding body makes with the speech she cannot provide. Cynthia Marshall notes the importance of this pointing for Marcus in the moment of his encounter with Lavinia: 'Narrative ordering must occur for Lavinia's plight to become the occasion for sympathy. Without a narrative framework, the figure of a mutilated woman in extreme circumstances functions ambiguously, as if oscillating between the possibilities of sensual appeal, ecstatic self-abnegation [the martyr function], shamed self-effacement, and imploring neediness' (2002, p. 127).

But a book Lavinia cannot be, despite the earnest, anxious wishes of characters and critics alike, precisely because her damaged body makes a shocking spectacle of ambiguity, miscomprehension, of exegesis gone awry. She has literally no words, no source of language. She is text's stunning failure. Aebischer helpfully puts this failure in theatrical terms: Lavinia, on the page, is silence from the moment she loses her limbs, and yet, in performance, that silence

> is made up for [...] by the actor. Whereas in the study, reading *Titus Andronicus* means reading Titus' grief in response to the textual gap left by his daughter's violation, in the theatre, the mutilated rape victim is insistently kept before the audience's eyes for six scenes. The actor's body *represents* the absence of words. Watching *Titus Andronicus* therefore means watching Lavinia.
> (Aebischer, 2004, p. 26)

Aebischer is interested in a rather different kind of return than the one I am exploring here. She wants Lavinia's battered and bruised body itself to return, finally to emerge from Shakespeare's margins and assume centre stage in the drama of her own undoing. Politically, I am in sympathy with this goal, and my readings of the play in performance will demonstrate the extent to which I find the reconfiguration of Lavinia's body a necessary feminist gesture as we work toward an ethics of violence against women in performance. But I also find this goal somewhat unsatisfying, because it assumes that simply by returning Lavinia's personal story to public performance space the problem of rape's effacement, its historical legacy within public space, and its potential impact on contemporary performance can be solved. To see the missing body returned, however, does not guarantee that we see what lies behind it, the elisions and evasions that have informed, and continue to inform, its cultural status as missed. If anything, the return of the missing body makes us feel better – now, how clearly we see! – and thus risks offering us incontrovertible proof of our power as eye witnesses at the expense of our obligation to witness beyond recognition, to witness the processes through which bodies and experiences go missing, to witness the cultural – and theatrical – power of absence and disappearance. What if the goal of a feminist rendering of *Titus Andronicus* was not to let us see clearly, to 'get' the missing story, but rather to show us the trace of the absence that haunts *Titus*'s stage, the limit of our ability to make sense of, and take ownership over, the ostensibly simple story Lavinia's cryptic body hides?

For a textual scholar of *Titus Andronicus*, Lavinia might be a body lost to troping, an unstable signifier, or an aesthetic object struggling to claim status as the author of her own story. For a feminist scholar, Lavinia might be a victim of violence whose voice and experience must be reclaimed in order for the play to have merit. But for a scholar of performance theory, Lavinia is foremost an actor in a play that goes awry when she loses her hands and tongue, the means by which all performers participate in the uniquely aural-gestural economy of the theatre. Watching *Titus Andronicus* does not just mean 'watching Lavinia' (Aebischer, 2004, p. 26): it means watching Lavinia *fail to perform*, watching her literally derail the play. Similarly, the moment of the long-delayed return – during which she inscribes Demetrius's and Chiron's names on the ground alongside the word 'stuprum' – is not just a matter of reading and writing, proof of Lavinia's essential connection with text. It is a *performance* of writing, an active, painful, full-body gesture that rehearses Lavinia's violation not by way of any well-worn convention dictated by the terms of the hue and cry, but by way of crude mime, the awkward, struggling contortions and oral invasiveness of forced sexual encounter. The repressed returns with a jolt as Lavinia drags Marcus's staff through the dirt, propped by her arms and mouth. She marshals the comparative clarity of the written word and replaces the 'map of woe' (3.2.12) she had become with the *image* of the signifier (two names plus a deed), neatly summing up her suffering by folding language into image at a staff's length from her body. A near literal re-enactment of the deed to which it is only supposed to gesture in discrete symbol and pre-screened language,[8] Lavinia's long-awaited metatheatrical return puns on its own formal process, riffing almost parodically on how the sanitized strategy of show-and-tell turns an externalized abstraction of rape into rape's ocular proof. Lavinia acts out her undisclosed scene of suffering in a manner that is without question more risqué than *The Lawes Resolutions of Womens Rights* would recommend; nevertheless, she manages to provide her male relatives with the information they need at last. However traumatic this rehearsal, it safely returns Lavinia to her father's mimetic fold, and dissolves her hysterical stage.

Lavinia's hysterical stage is the one onto which she steps, bloodied and torn, at the end of Act Two. Marcus discovers her and promptly invokes the story of Philomela and Tereus (2.3.26) in such a manner that we should be certain he has already guessed the source of Lavinia's suffering. But no sooner does Lucius ask Marcus to explain the sight of her than Marcus is unable to comply (3.1.88–91). The Andronicii seem unable, or unwilling, to recognize the rape, despite the entirely straightforward, if quite grotesque, spectacle of shamed modesty Lavinia makes. Their third-act confusion has generated a fair amount of speculation among critics and indeed does pose several vexing questions. Why does Marcus speak of Philomela and then promptly forget the connection? Why doesn't Lavinia's body, heavily troped for us by her appearance and by Marcus's blazon, function according to convention? Emily Detmer-Goebel (2001) argues that the confusion over Lavinia's violation emphasizes the need for Lavinia's own testimony in making rape known; Marcus can offer a revealing soliloquy for audience benefit, but until Lavinia finds a way to tell him of her experience – in other words, to raise the hue and cry – he cannot know or make known her experience within the world of the play, and he cannot obtain justice. As an extension of this sensible and historically sensitive argument, I propose that the problem is not that the Andronicii do not understand that Lavinia has been raped, but rather that, because they do not yet know *who* is responsible, they are unable to conceive of the rape as an assault on their own bodies (or their family body) by another man and hence they are unable to conceive of the rape *as rape* at all. Lucius's anxious question (on 3.1.88) isn't 'what happened to Lavinia?' but rather 'who did this to Lavinia?' – until the latter is answered, the former is of little matter to anyone. This is the problem that the hue and cry is meant to address: Lavinia's job is to show and tell her pain in order that the perpetrators may become known and the event may be translated into an experience of and for her male relations. Until that happens Titus's revenge drama cannot continue. The confusion and anxiety Lavinia's body incites thus has less to do with what it may symbolically reveal than with its performative limits, which in turn mark the limits of its ability to return evidence: the bloody stumps where Lavinia's hands and tongue used to be speak of her as Philomela *in extremis*, but they also prevent her from accusing her attackers, and thus as indices of *her family's* experience of the rape they are incomplete. Her audience can only watch in distress as a seemingly clear picture of her suffering dissolves, leaving emptiness in place of a surfeit of symbol.

On a hysterical stage meaning slips sideways: the hysteric's words collide with the literal, abrogate the distance between signifier and signified, turn the referent uncanny, hauntingly (un)familiar. Elin Diamond reads Julia Kristeva's feminist hysteria as a kind of mimicry, a logical extension of the transgressive mimetic power articulated by Luce Irigaray in her reading of Plato. The hysterical performer does not act: she does not make motions that correspond to *a priori* meanings, let alone seamlessly ape well-worn conventions. Instead, she thwarts word and gesture, signs missingness, and traps her interpreters within it. Act Three of *Titus Andronicus* sets a scene in which Lavinia's trauma can be recognized, but cannot, for lack of an accused, for lack of the means to express accusation, yet be translated out of her body. In this hiatus space Lavinia's trauma can only be about her, within her: it has been frozen in the moment when it is meant to become public – the moment of show-and-tell, of the metatheatrical return – but cannot because she cannot act, cannot complete the performative transfer to patriarchal space. Although her mutilated appearance causes Lucius to insist 'this object kills *me*' (3.1.65, my emphasis), in her not-quite-fallen state Lavinia cannot be co-opted. She is less an emblem of castration ('such a sight will blind a father's eye' [2.3.53]) than she is an echo of Freud's mythical castrated woman, suspended in the lost moment of her never-acknowledged violation, before it becomes a mirror of male subject-formation.

For Freud, castration fear marks the psychic subject's entry into culture, when he (always *he* in this case) is forced to confront his vulnerability, the ruse of his invincibility, and yet is compelled by his recognition of that vulnerability to imagine himself more powerful than ever in a kind of talismanic proof to the contrary. As Freud's story goes,[9] the boy-child witnesses his mother's lack of a penis, imagines her to have suffered a violent punishment that might also fall upon him, and dissolves his Oedipal complex by projecting his fleeting image of her long-lost castration onto his own body. The notion of what might have happened to *her* never crosses Freud's mind except as a function of what might happen to him. Her suffering, even more theoretically far-fetched than his, is without meaning within Freud's psychoanalytic framework. The boy's symptom (like the fetishist's obsession) marks that suffering's uncanny, metapsychical return. Lavinia's unmediated violation is just such an uncanny castration terror: a spectre of violence that refuses transference into the realm of male psychic and social power. Her obscure grunts and flailing gestures collapse into her damaged body and cannot traverse the distance between her experience and that of her family, cannot signify within the limits of their imagining. Hence young Lucius's terrorized cry: 'I know not what you mean' (4.1.4), and Titus's fraught, anxious reply: 'Fear her not, Lucius – *somewhat* doth she mean' (4.1.9, my emphasis).

A body in violence to which no stable homosocial assignation can be made, Lavinia throws the stage into disarray. The Andronicii, in turn, attempt to recuperate their damaged mimesis by competing zealously for the privilege of echoing Lavinia's trauma on their own bodies during Aaron's macabre hand comedy (3.1.151–206). Later, fearing their entrapment in 'dumb shows' (3.1.132), Titus takes matters into his own hands and adopts the role of director. He lectures the company assembled at dinner 'on the proper theatrical gestures to express outwardly their passion' (Hulse, 1979, p. 114) and casts Lavinia's continued suffering directly in terms of her inability to act it out (to purge it by performing): 'Thou map of woe, that thus dost talk in signs, / When thy poor heart beats with outrageous beating, / Thou canst not strike it thus to make it still' (3.2.12–14). To the heels of this instruction Titus attaches another, much more sinister: 'Wound it with sighing, girl, kill it with groans, / Or get some little knife between thy teeth / And just against thy heart make thou a hole' (ll.15–17). Marcus's protests at this suggestion (ll. 21–2), though they may echo early modern England's own discomfort with suicide (including the self-immolation of violated women), do not grasp the essence of Titus's instruction. He is less teaching Lavinia about suicide than he is trying to teach her how she might, despite her limitations, play Lucrece instead of Philomela and bring her performance of the ravished heroine to its inevitable conclusion. Her death may then prove beyond doubt her non-consent and complete the transfer of suffering from her body, moved beyond its misery, to his, left to the rites of vengeance and mourning.

Titus's backstage instruction might be suitable for the doomed classical heroine and her histrionic followers on the early modern stage, but it offers little help to the more mundane victim. For the latter, the play turns to Marcus. In 4.1 he borrows the role of director from Titus and shows Lavinia how, by taking his staff in her mouth while guiding it with her severed limbs, she can provide her relations with the information they need. The specifics of Marcus's suggestion are notable not only because they initiate the crudely literal rehearsal of Lavinia's rape that I describe above, but also because they are designed to fill her empty mouth, replace her missing hands – to provide her with the prosthetics of performance. As Lavinia shows-to-tell her tale of woe she becomes once more a fully functioning actor within her family's drama, taking up her (appropriately subordinate) role within Titus's revenge scheme.

The staff-in-mouth rehearsal of Lavinia's rape works to resuscitate Titus's wounded mimesis, but does it succeed? Certainly, relief floods the stage; Titus goes on to plan an elaborate scene of vengeance in which Lavinia will be Virginia to his Virginius in one last attempt to suture her damaged body with the familiar pleasures of known narrative. (In the space of half a play Lavinia has added no less than three popular raped heroines – Lucrece, Philomela, Virginia – to her repertoire. If she was once unnervingly empty, she is now an excessive theatrical sign.) Yet the moment of rehearsal itself remains problematic. In contrast to the metonymic frame governing rape's revelation in *Lawes Resolutions* – where bodily injury and woeful lamentation are intended to stand as associative substitutes for the missing act – *Titus Andronicus'* show-and-tell is scripted around a much more literal process of substitution: word becomes image, staff becomes phallus, mouth becomes a *vagina dentata*, and Lavinia emerges as her own ravisher. The moment thus (unwittingly?) stages the deep prejudices, assumptions, and expectations governing the set conventions of the metatheatrical return – as well as the limited usefulness of those conventions when faced with a rape victim's extraordinary bodily trauma. The risk assumed by such a staging is exactly the risk posed by staging rape itself, unmediated by figuration: that it may draw undue attention to the substitutive quality of theatrical process, to the fact that the original act to which Lavinia's rehearsal refers is in fact no act at all, but a gap in the iconography of the play. Lavinia is not simply re-enacting her rape; the actor playing Lavinia is creating it for the first time. His/her performance has no precedent, aside from the displaced gestures of Marcus in his tutelage. Lavinia's Act Four 'act' performs rape as an essentially, uncannily theatrical matter: her copy has no model, is a copy of Marcus's copy of nothing but the blank space between scenes.

Lavinia's queer return does not, of course, suggest that rape is somehow merely representational, not 'real'; rather, it allows *Titus Andronicus* to reveal how rape's theatricalization *becomes* its public reality at this moment in time, and to demonstrate how essential that theatricalization is to the return of an ordered homosocial space in the late sixteenth century. The play stages many of the anxieties about rape's epistemology circulating in early modern culture by foregrounding a parallel anxiety about Lavinia's status as a performer – one charged with reproducing her experience for her male protectors in such a fashion that it becomes, seamlessly, about them. Instead of returning her pain to their space, Lavinia reveals what the metatheatrical return usually masks: that rape can be neither so easily known nor so easily co-opted; that, observing from the vantage of the social, we will never be able adequately to witness sexual violence, know all its detail, the true extent of its consequences.

Spectacles of absence

If *Titus Andronicus* is less a spectacle of grotesque violence than a spectacle of rape's uncanny absence, how do we mark this play for feminist revival? How do we stage the representational politics that surround Lavinia's rape without rehearsing the erasure of embodied experience the metatheatrical return works to accomplish? How do we bear witness to Lavinia's suffering while also bearing witness to the limits of our recognition, to the impossibility of our ever seeing or knowing that suffering intimately or completely? A tall order indeed: how do we offer Lavinia subjecthood in performance without essentializing her experience of rape and our access to it – without forgetting that Lavinia is not just a subject or a victim but also a performer?

In 1987, Deborah Warner directed *Titus Andronicus* for the Royal Shakespeare Company in a widely lauded, groundbreaking production that presented the rape and its aftermath as an experience of profound personal trauma for Lavinia (Dessen, 1989, pp. 51–69; Bate, 1995, pp. 62–9).[10] Unlike earlier noteworthy productions, this one not only took Lavinia's rape seriously as a central event in the play (Aebischer, 2004, p. 42) but also worked hard to contextualize the rape's political meanings, focusing attention on Rome's explicitly patriarchal social order as 'one of the material manifestations of Titus's absolute authority' (MacDonald, 1993, p. 196). Gender was power in this production, and Lavinia became the vehicle through which Warner made the implications of this equation for women crystal clear. Sonia Ritter's Lavinia was endowed from the start with a sense of herself as a desiring subject; she took pleasure in her relationship with Bassianus, her new husband, and, after being cornered by Demetrius, Chiron, and Tamora in the forest just prior to the rape, her pleas to Tamora for mercy were both heartfelt and tinged with rough rage. As she begged Tamora, weeping and clinging, she nevertheless conveyed a sense of her anger's power, an edge her performance retained even as Chiron and Demetrius dragged her, shrieking, away. This aspect of Ritter's work had a decidedly 1980s feminist quality: as Aebischer remarks, she 'did not allow viewers to establish any easy equivalence between femininity and victimisation but rather stressed the desire and ability of the traumatised rape victim to survive and seek revenge' (2004, p. 45).

After the rape, the tenor of Ritter's performance changed dramatically. Demetrius (Piers Ibbotson) and Chiron (Richard McCabe) re-entered giggling, crawling, lewdly mocking Lavinia's agony with their bodies. Lavinia, meanwhile, appeared in a very low spotlight, walking extremely slowly; the light focused tightly on her as she struggled to move forward. Every muscle in Ritter's body signalled that Lavinia had been physically and emotionally destroyed; as she reached centre stage she fainted backward, collapsing into what eventually became a kind of foetal curl. When Marcus (Donald Sumpter) entered she rushed downstage into the shadows, gesturing wildly; she remained in low light throughout his lament, miming her distress, finally collapsing into his arms. Ritter's physicality was designed to foreground the material reality of Lavinia's bodily and emotional suffering at every turn, the very thing for which the metatheatrical return cannot fully account. In addition to her performance of physical trauma, she also offered numerous suggestions that Lavinia was suffering from a form of post-traumatic stress disorder, if not rape trauma syndrome. These mental cues allowed the production to challenge easily naturalized links between madness and femininity, forcing audiences to confront the causal connection between violence against a woman and her ensuing emotional dysphoria.[11]

While Ritter's intense focus on Lavinia's embodied experience of rape had feminist merit in itself, its most striking effects emerged during Act Three. Despite the overwhelming spectacle of suffering produced by Ritter's increasingly catatonic body, Marcus, Lucius (Derek Hutchinson), and especially Titus (Brian Cox) made short work of ignoring her utterly, to often absurd effects that highlighted the disconnect between

Lavinia's experience of her trauma and her family's understanding of it. When Marcus and Lavinia entered to Titus at the top of Act Three, she remained in dim light while Titus expressed his pain and self-pity from the brightest spot at centre stage. The technical contrast underscored the textual one: Titus is the centre of this scene, his suffering the suffering that matters. It also, however, showed up the important performative implications this moment embeds: Titus, who can stand and play the patriarch, is given pride of place on stage, while Lavinia's broken body, unwilling or unable to act on cue, must be stashed in the wings, nearly out of sight.

Warner's feminist intervention was for me most remarkable right here, in her staging of the men's surprisingly easy, dramatically seamless forgetting of Lavinia's shattered self. If Ritter's Lavinia had been so traumatized by her rape as to prevent her from continuing with her role, the show would nevertheless go on without her. Eager to comfort, Lucius takes Lavinia in his arms; he lets her go dismissively and she heaves over, sitting up only very slowly. Cox plays Titus as Lear, brimming with self-assured authority and self-pitying lamentation; he takes hold of Lavinia to crown his sorrowful speech, but when Lucius arrives to comfort him he quickly lets go and she collapses on the floor. A funny moment, and yet there's no laughing matter here: the peculiar absurdities of Act Three took on a darkly comic edge as the men, unable to read Lavinia's body via the framework of the return, worked overtime to disavow its troublingly unmediated presence in their space. Ritter was collapsed on the floor at centre stage when Aaron (Peter Polycarpou) entered to propose the trade of a hand for Titus's sons' lives; he stood beside her collapsed body and yet garnered alone the Andronicii's attention as he offered them a way to absorb their family's suffering into their own bodies. While the men struggled over who would give his hand, Ritter remained on the floor, writhing in her private agony; her soiled but still bright yellow-gold dress made her plainly visible among the otherwise dull wooden set and other costumes, calling unnerving attention to the men's desperate need to forget her body for their own, for the family body, and for the body politic. Finally, after writing Demetrius and Chiron's names on the ground (with Marcus's staff between her stumps and against her shoulder, not in her mouth), Ritter collapsed once more; then, propped upright and facing the audience, she fell into a doll-like stillness, prefiguring the broken puppet she would become when Titus snapped her neck as she sat on his lap in the play's final moments. In the aftermath of what had quite plainly been for her a second rape Ritter's catatonia grew increasingly pronounced, and in the shadow of this ever-more virulent trauma the men's refusal to engage with her became overtly political. With its focus on 'the social and political relationships in which the text grounds its savageries' (MacDonald, 1993, p. 199) and its minimalist, mud-coloured set, Warner's production offered plenty of Brechtian echoes, and Ritter's Lavinia found her Brechtian *gestus* – the action or gesture that both explains, but also exceeds, the meaning of the play – in repeated acts of collapse. When her relations no longer bothered to prop her up, she simply fell down; when Titus cut off his hand, she rose jerkingly, flew toward him, and then collapsed once more – as though she had, perversely, taken *his* suffering into *her* body. Later, she took his hand not between her teeth but in her arms, hugging it as though it were the site and source of her own loss, something to which she, in her damaged and keening state, had no other access. Ritter's collapses registered again and again as morbidly ironic, Lavinia's relations stepping around her even while she lay at centre stage; they also, however, registered as marks of Lavinia's catastrophic loss, a loss that had somehow left her stranded in an unconscious elsewhere. They made me laugh at the men's myopia, but they also reminded me that Lavinia's physical expressions of her suffering were finally expressions of its inaccessibility – for her father, uncle, and brother, for her audience, and for herself. And yet, even in this acknowledgement I wondered: when we watch Ritter suffer and fall down, do we really note Lavinia's loss as loss, her pain as disconnection, as distanced from itself? Or are we encouraged to see something else in place of that disquieting emptiness?

To a 'politically engaged' spectator (Barker, 2007), Ritter's Lavinia was less a body than a ghost in Warner's production. She was the lamentable object around which Titus, Lucius, and Marcus organized their performances, but at which they seemed unable to look directly. Left on the stage's margins, Ritter's ever-collapsing body charged viewers to see beyond conventional acts of spectatorship and become true witnesses: to pull our attention from Brian Cox's outlandishly tragic Titus, acknowledge Lavinia's obscure, often impenetrable actions. and begin to encounter the magnitude of their import. But, in the nature of this very call to witness, problems arose. As Ritter played the on-stage spectre, her devastated, ever-moving body simultaneously drew spectators' eyes to a performance of trauma that was designed to be literally painful to watch: Warner sought 'ways of making [Lavinia's experience] unbearable' for audiences (qtd in Aebischer, 2004, p. 61). Prompting us to feel Lavinia's pain is, to be sure, a welcome alternative to refusing that pain in performance, but such a gesture also risks, even encourages, pain's affective co-optation, its transfer from Lavinia's mind and body not onto the stage but into the auditorium. Despite the tremendous critical potential of Lavinia's spectacular forgetting-in-plain-sight, Warner's ultimate message was not that Lavinia and her trauma had been missed and would continue to go missing – that her bodily experience defies our access even as her family demands her disembodiment – but that what was missed of Lavinia's experience was missed *by those on stage*, while offstage witnesses could claim the privilege of both knowing her more intimately and reading her 'martyr'd signs' more clearly. At the Act Three supper table, Ritter and Cox played up Titus's incomprehension of Lavinia's gestures even as their meanings became more and more painfully obvious. Clearly desperate for water, Ritter mimed her thirst unmistakably but got no help from the others at the table as they 'struggled unsuccessfully to read her signs' (MacDonald, 1993, p. 199). During the performance I watched on archive video, the audience laughed loudly when Cox insisted on his ability to 'interpret all her martyr'd signs'; the comical disconnect between Titus's authoritative claim and Lavinia's obscured meaning was clearly one of the main points of this scene (see Figure 1), and one given traction only because it appealed to audience members as the true authorities on Lavinia's hampered language.

While I do not wish to generalize an audience response to Warner's production either from one recorded performance or from printed reviews in the mainstream press, both offer strong hints about what aspects of the production were ultimately considered memorable. Overwhelmingly, the British press understood Brian Cox's Titus as the star of this show, both during its initial 1987 Swan Theatre run and again at the Barbican Pit in 1988. Reviewers repeatedly called Cox's performance a 'trial-run' for an eventual turn as Lear (Billington, 1987, p. 636; see also Gardner, 1987; Gordon, 1987; Billington, 1988; Hiley, 1988; Morley, 1988; and Spencer, 1988) and gave the majority space in their reviews to assessing the nuances of his performance. With only a pair of notable exceptions (Edwards, 1988; Marmion, 1988), Ritter's performance is, *pace* Lavinia's body, an afterthought in these reviews: almost universally praised, but praised in a sentence or less when it is mentioned at all. Those who do mention Ritter code her work as successful because movingly realistic but also helpfully subordinate. Reviewers speak of her as 'heartbreaking' (Morley), 'unbearably affecting' (Spencer), 'the true tragedy of virtue' (McAfee, 1988); they also quite easily assimilate the many moments of dissociating uncertainty in Ritter's performance into a coherent sound-bite, reading in

Figure 1 Brian Cox as Titus and Sonia Ritter as Lavinia in Deborah Warner's 1987 production of *Titus Andronicus* at the Swan Theatre. Photo by permission of the Joe Cocks Studio Collection, Shakespeare Birthplace Trust.

her work just enough pathos to undergird Cox's own tragic *tour-de-force*. (Michael Billington even went so far as to call Cox's Titus 'a concerned father' [1987, p. 637]!) Jack Tinker's 1988 review sums this trend up perfectly: 'As a study of a good man turned inhuman by the inhumanity of others, the production [...] is possessed of a timely moral force. Who can look on Sonia Ritter's quivering Lavinia, her tongue, hands and chastity ripped from her, and not know pity?'

Warner's reviewers, like the Andronicii themselves, seemed to ache for the metatheatrical return. They read in Ritter's performance a Lavinia who functioned as the embodiment of a particular theatrical convention: one that moved them to pity and fear, and to praise her affecting dramatic presence, but one that did not require them to acknowledge the naked way in which Ritter enacted an experience that was continuously marginalized on stage, deliberately missed in the whorl of Titus's Lear-like declamations. Warner sought our visceral witness by assaulting viewers with the raw power of Lavinia's damaged body, but in the end she risked earning too much of our self-satisfied ownership over Lavinia's pain because Ritter's performance conformed, in so many ways, to late twentieth-century assumptions about what trauma 'really' looks like. Ironically, because Ritter and Warner chose to stage Lavinia's body as psychologically 'realistic', reviewers were able to slot her far too easily into the framework through which they read the production as a whole. Warner did a remarkable job of framing the experience of women's bodily and psychic suffering as an echoing loss within a clearly demarcated patriarchal territory, but her production ultimately failed to politicize its larger cultural elision in a way that might disorient audience expectations about what the experience of sexual violation should look like, for what (or whose) purpose it should be performed, and to whom it grants the power of complete understanding.

What Warner's production finally missed – even as it performed a woman's suffering as real, embodied, all-pervasive, and ultimately forgotten – was the opportunity to engage that suffering as a function of its representational history. The men on stage very much missed the point of Lavinia's trauma, but they appeared to be the only ones to do so; moreover, in the end, their glaring miss does not seem to have affected the reception of their heroic performances. I want to argue now that a critical feminist response to *Titus Andronicus* needs to deal with rape's early modern representational history – with the pernicious and pervasive logic of the metatheatrical return as it organizes what we see and understand of Lavinia not simply at the level of character development but more broadly and deeply at 'the level of the signifier' (Belsey, 1999, p. 5). A feminist staging of the play does not need to be earnest with Lavinia, but it should be responsive to and ingenious with its reproduction of the theatrical trap into which her rape plunges her. Julie Taymor's film version of the play, *Titus* (1999), though neither overtly feminist nor particularly interested in Lavinia's rape as a centring principle, comes closer than Warner's to managing what I envision as a feminist performance of early modern sexual violence. It articulates and exploits rape as a historicized act as well as an ethical challenge constituted by its own theatrical double edge: its notorious refusal to appear coupled with its crude, even shameless specularity.

Titus is based directly on Taymor's 1994 Off-Broadway production of the play, notable for its 'Penny Arcade Nightmares' and other gruesome vaudevillian touches; it also borrows heavily from previous Shakespeares on film (most obviously Luhrmann's *Romeo + Juliet*, with its heavy glam-rock quotient and self-referential 'red curtain' motif), the play's own stage history, as well as a variety of cross-cultural theatrical traditions (Aebischer, 2001). This patchwork legacy, coupled with the sleazy fairground quality Taymor evokes in many of the film's cruellest moments, makes clear that *Titus* understands theatricality, twinned with bloodshed, to be the play's major structuring principle. With its catalogue of horrors, of which Lavinia's rape is finally only one, *Titus Andronicus* is a tight-rope walk on the thin red line between violence as ritual and violence as ritual *entertainment*; between violence as a sacrifice to sublimated terror and violence as titillation, the perverse production of sublimating pleasure. Taymor, in turn, transforms this cathexis of violence and pleasure into a multi-faceted audio-visual signifier on the border violence and pleasure share with the theatre (L. Hopkins, 2003a, p. 65).

Taymor frames the play's collapse of violence into theatricality with theatre's encounter with the Real: the moment when play becomes risk and then disaster, the moment when terror crashes in on performance and throws ethical choice into relief. *Titus* opens in a twentieth-century kitchen, where the boy who will later be young Lucius (Osheen Jones) eats his lunch while playing war games with his toys. His play becomes violent, and in the moment it loses control *Titus Andronicus* crashes in. Young Lucius is swept off to the coliseum where Titus's troops are returning from battle in formation; they are at once Roman warriors, chorus-line dancers, and video-game soldiers, marching in stylized formation. This trans-historical, trans-generic playfulness defines Taymor's aesthetic, but it does not imply solipsistic self-referentiality: as Clara Escoda Agustí writes, 'Taymor creates spaces for self-reflexivity and uncanny recognition, interstices of thought whereby the audience may question the kind of ideology and gaze it is participating in' (2006, p. 1). The story of the film is in many ways the story of young Lucius, of his journey through and, finally, away from the oblivious cruelty of violence-as-play. Taymor opens as he tortures his toys in a familiar domestic space, tracks him back in time through his generous care of Lavinia once she becomes a macabre stand-in for his long-abandoned playthings, and closes on a sentimental image of him carrying Aaron's baby into the sunset. The film ends in an uncertain future but transports us back to our beginning, to Jody Enders and Mrs Coton, relentlessly asking the all-important question: when does theatre cross an ethical line, and, once it does, how will we respond?

The film's final scene produces perhaps its most startling image of Taymor's ethics. One moment we are in Titus's dining room, watching Tamora eat her children; the next, we have been returned to the coliseum, the set for which was an extant Roman ruin in contemporary Croatia populated by Croatian extras playing spectators to the scene (L. Hopkins, 2003a, p. 64). The horrors of this ending are, manifestly, real – for the Balkan extras much more so than for English or North American audiences in 1999 – but they are also clearly performed, a play on and with violence that skirts a line far too easy to cross. Thousands of onlookers stare down on a film set that has just become a stage set dwarfed by the mammoth open-air auditorium. They sit in stunned silence. Perhaps that silence is in the script; or, perhaps they have seen this all before, know there is nothing to say. As Courtney Lehmann, Bryan Reynolds and Lisa Starks write, 'Taymor's film redefines the subjective territory of spectatorship as a perilous negotiation of, and ultimately a choice between, alliance-building and terrorism, accountability and complicity' (2003, p. 223).

Given the extremes of Taymor's stylization elsewhere, Lavinia's first encounter with her rapists is oddly realist. The forest is a forest; Lavinia's (Laura Fraser's) dress and that of Tamora (Jessica Lange) and her sons (Jonathan Rhys Meyers; Matthew Rhys) are largely historically consistent. But when Lavinia returns post-rape, the tables have turned. She lands in a post-apocalyptic swamp punctuated by gnarled trees; she awaits discovery atop a tree stump – a blatant reference to her own stumps, which have been costumed to end in branches. The branches make a queer echo to text as well as image when Marcus (Colm Feore) eulogizes 'her two branches, those sweet ornaments / Whose circling shadows kings have sought to sleep in' (2.3.18–19). At this point in the play Lavinia is typically a visual negative, a body emptied of signifying power and as such a source of powerful anxiety; Taymor casts her instead as a literal sign of Marcus's text, of her own (scripted) sorrow. A hysterical symbol, Fraser's Lavinia is simultaneously sign and referent, morbidly, uncannily sufficient in herself. She moves slowly, but not without elegance. Framed for our consumption, she seems at first charming despite her obvious trauma, but when her mouth opens and blood spills out in a long, slow arc, the spell breaks and Lavinia becomes the image of a Fury.

For Fraser's Lavinia, unlike for Ritter's, the trauma of rape marks a deliberate shift *away* from realism. Young Lucius visits a doll-maker and purchases two prosthetics that give Lavinia the quality of a living puppet – but after a fashion very different from Ritter's turn at puppetry. In Warner's production, Lavinia's awkward movements and doll-like blankness serve to underscore the material and psychic reality of her trauma; in Taymor's film, they are marks of her theatrical self-consciousness, her performer's status. Feore's Marcus shows her how to tell her woe, taking his staff in his mouth and between his arms as he transforms it into a pen; Fraser, however, refuses the textual implications, as well as the obvious payoff of direct substitution the lesson implies. She looks long and hard at the staff's crown (shaped to be, unquestioningly, a wooden penis) and rejects it – *vomits* it, in fact, after almost but not quite swallowing it. Pressing the staff instead into the crook of her neck like a violin, she roars into action accompanied by a hard-core metal track – the same track that is associated with Demetrius and Chiron elsewhere in the film.[12] The effect is an extraordinary audio/visual dissonance: the rough music is now painfully Lavinia's, and it seems to fly from her manic bow. Taymor meanwhile crosscuts Lavinia's speech/act with a stylized, blue-washed flashback sequence that turns her show-and-tell into a music video. Lavinia is the centrepiece, costumed as she was in the swamp but with a hint of the imperilled vixen; though she wears a look of vulnerable terror she channels Marilyn Monroe, her skirt flying above her waist, her lopped arms grimly inadequate to keep it down. Demetrius and Chiron flash in and out in profile, looking vicious, demented; meanwhile, tigers attack a young doe to complete the image allegorically. Set to the pounding beat, the clips compound the performative dissonance already at work within the main action of the scene.

Several critics have expressed concerns about this jump-cut representation of Lavinia's metatheatrical return, and not without good reason. Just as Warner risks over-privileging audience access to Ritter's trauma, Taymor risks pandering to an oversimplified audience pleasure in Fraser's Marilyn-esque rock video. Pascale Aebischer writes: 'Disturbingly, the association with Monroe codes Lavinia both as a victim of male exploitation and as "available" and "asking for it"' (2004, p. 47). Roberta Barker frames the same worry in a broader representational history: 'Hollywood goddess, religious icon, bride, willing sacrifice: all of these images fix Lavinia in [a] timeless, spaceless realm [...] She becomes, not a human being who has suffered physical and psychological abuse, but an "idea" of the universal feminine' (2007,

p. 104). David McCandless, meanwhile, compares *Titus* unfavourably to Taymor's 1994 Off-Broadway production of the play, arguing that the latter's 'Penny Arcade Nightmares' cannot be translated onto film without reifying the cheap titillation they were designed originally to critique. McCandless reads Lavinia's tell-all as a 'surreal' projection from her 'battered' mind, inexplicably overwrought as her own traumatic recollection, but also inexplicable to a film audience without the Brechtian frame of reference Taymor's carnivalesque ethos lent the original stage version (and which the video reproduces almost image-for-image but far more sleekly) (2002, pp. 501–2).

These critiques of Fraser's rock-star moment ably set out the representational risks Taymor takes with *Titus* in general, and Lavinia in particular, but they do not fully account for that moment's engagement with the terms of performance on which Taymor builds the rest of her film – and, indeed, the rest of this scene.[13] Fraser's Lavinia is, as I noted above, no realist: she doesn't model herself on a particular damaged heroine, but rather assimilates a host of references, from Grace Kelly and Marilyn Monroe to hints of Vivian Leigh in the 1955 Brook/Olivier production of the play. Overcoded by a history of feminine celebrity, Lavinia becomes the copy of a copy of the blank between performances of demure suffering. And, so rendered, she (not surprisingly) finds herself thrust into the TV camera. In her 1994 stage production, Taymor shifted rape's revelation into a vicious, distorted vaudeville: standing exposed on a box, lunged at by men made up to be tigers (McCandless, 2002, pp. 494–5), Lavinia literally became the freak show required by the hue and cry. On film, Taymor generates an equivalent shift by upping the production values and transporting Lavinia into a much more contemporary circus: mainstream pop video, the spectacle of innocence at the end of the twentieth century. Lavinia becomes Britney Spears circa 1999: the signifiers of chastity are always in question, though the performance of chastity, like the hint of danger, remains picture-perfect.[14] We learn the identity of her aggressors without having to witness anything unseemly, and without her having to speak of their heinous deeds; reference to the tiger gives us a familiar iconography.[15] The loud music, blue wash, and windswept look give the clip the air of MTV cool; this is the representation we've all, supposedly, been craving.

Meanwhile, on the other side of the celluloid negative, Lavinia in real time crafts a different kind of show-and-tell, a rebellious response to the music-video version of her story required by her official witnesses. Refusing to take Marcus's staff in her mouth, she refuses what the video embraces – the carefully wallpapered money shot, the 'payoff' (Phelan, 1993, p. 19) of a metatheatrical return based upon conventional reality-effects and pervasive expectations about who she is and what her experience ought to have been. Working against the substitutive equation staff/penis, Lavinia displaces her real-time act into a gesture that deliberately *does not* reference forced intercourse, hints at nothing remotely sexual, is rather a kind of solo improvisation that allows Lavinia to release some of her own pent-up anguish along with the information (and its attendant catharsis) her family has been seeking. This version of Lavinia's telling is neither crude vaudeville nor high-tech video schmaltz, but a jarring, straining dance. She dances against the ravished and vulnerable, ballet-beautiful icon from the swamp, against the sexually provocative figure in the video, against the sexual prejudices that frame her woman's body and its experience of violence, against the performance frames designed to contain and perpetuate those prejudices. She dances in order to express in and through her maligned body something of the fractious, erratic, sensory-overloaded quality of her violation. Rather than embrace the tools of Marcus's patriarchal mimesis, Lavinia displaces signifier (staff/pen) from signified (penis/rape), and crafts the telling of a more complex story, woven through her body into a dissonant, cacophonous, raging music.

When Lavinia confronts Marcus's staff, theatre crashes in. The 'video' tries to keep her representation in line, but Lavinia will take to a different stage. The moment of show-and-tell, as I have been arguing, is the moment when theatre attempts to cross a line, to frame and contain an experience that cannot be so easily boxed into a proscenium. It is the moment – to invoke the questions with which I began this chapter – when the metatheatrical return poses risk and threat alongside the (admittedly often dim) hope of justice, but it is also the moment through which we may frame an ethical performance response to that threat, a respectful response to the disorienting terrors of sexual violence. Such a response would make space for a spectator to become a witness to Lavinia's experience – not just to her experience as a rape victim or as a feminist survivor, but to her experience as an actor in the long-running show of rape's performative effacement, to her experience as a victim of rape *and* of rape's pernicious, malicious representational history. In Taymor's film, the moment of show-and-tell responds directly to this history as it becomes a site of ethical choice for the performer in Lavinia: she must decide how she will confront her violation, how she will make it public, make it social, give it a reality beyond her body. Choosing to do so *through* her body, in a series of gestures that resist transparency even as they reveal her attackers' names, Fraser's Lavinia reveals the return to be inherently plastic, malleable, open to a variety of rehearsal options. The metatheatrical return does not need to be a re-*enactment*; it can also be a re*creation* – or just plain creation. The return can take several forms: it can mimic expectations while meeting them, or it can mime expectations and produce something new, an expression of something not quite seen, as the residue of expectation. It can meet a social as well as a personal need; it can incarnate rape's extra-corporeal afterlife while also acknowledging that such an afterlife does not replace, negate, or overcome the complex, fractious, angry inward resonance of violation. Finally, in all these ways, it can mark the beginnings of a politically, historically, *and* representationally aware performance of sexual violence, one that acknowledges the material experience of raped women's suffering but does not stop there.

3
The Punitive Scene and the Performance of Salvation: Violence, the Flesh, and the Word

> [T]he symbol manifests itself first of all as the murder of the thing.
> (Lacan, 1968, p. 84)

The theatricalization of sexual violence works to absent the lived complexities of a survivor's suffering as it transforms rape into an experience proper to public space. And yet the conventions of rape's metatheatrical return, as we saw in Chapter 2, are pliable enough to be refashioned into a rebel performance of rape's constitutive invisibilities, provoking acts of witness 'beyond recognition' (Oliver, 2001). In this chapter, I turn away from sexual assault and toward the much more legally tricky notion of domestic violence against women in early modern England. Building on the forked logic of the metatheatrical return, I examine the relationship between the battery of wives in the late sixteenth and early seventeenth centuries and the acts of domestic theatre designed to keep that battery from appearing in public as anything more than a reasonable, indeed a *spiritual*, act of household care. What happens, I ask, when the language, tropes, and performance tools used to shape a woman's body in violence into a body in grace begin to break down? Is the 'performance of salvation', like the 'metatheatrical return', subject to a queer, unexpected haunting? What kinds of subversive cultural work might the battered female body do in early modern social space, and how can contemporary theatre artists harness its power?

Staging grace against the odds

Sometime in the middle of 1590, Katherine Stubbes, the 19-year-old wife of Phillip Stubbes, became pregnant with her first child.[1] While we have no historical evidence to tell us exactly how the pregnancy fared, we do know that Katherine, a devout Protestant, apparently suspected that childbirth would kill her. According to her husband Phillip, once her pregnancy became known Katherine told him, 'and many other her good neighbours and friends [...] not once, nor twice, but many times' that she would live to bear no more children (Stubbes, 1992, p. 144). Joan Larsen Klein suggests that Katherine actively willed her own death with this persistent foretelling (1992, p. 140); to her husband, though, Katherine's prophecy was proof of her enduring faith, a gift from God 'revealed unto her by [His] spirit' (Stubbes, 1992, p. 144).

Katherine's son was born. She healed well initially but then fell ill with fever and, perceiving death to be near, orchestrated a bedside spectacle of testimony and witness in order to confirm her earlier prediction. While my reference to 'spectacle' within this resolutely Protestant scene may seem anachronistic, many historians have noted that Protestant commitments to 'plain speaking' through the early days of the Reformation and counter-Reformation often borrowed known strategies of public performance, especially during interrogation or on the martyr's scaffold (see Knott, 1993, pp. 70, 79; Diehl, 1997, pp. 185–94; Burks, 2003, pp. 37, 42). In similar fashion, Katherine's sickbed became a stage (Phillippy, 2002, p. 93) on which she mounted an exemplary performance of her faith, underwritten by an equally impressive performance of personal erasure. If the absent star of Katherine's scene was her Heavenly Father, His unsung understudy was her suffering, aching, vulnerable body. Katherine was dying; there could be no question for anyone around her that she was in serious, visible pain. And yet, in his account of her death, Phillip Stubbes is careful to include references to Katherine's body only when it appears to be miraculously at peace despite all likely evidence to the contrary.[2]

According to Stubbes, shortly before her death Katherine fell into an 'ecstasy'; friends and relations gathered around her to grieve, pray, offer support. When she returned to herself she offered an extended sermon to the assembled crowd, whom she asked to 'be witnesses' to her Christian life, faith, and belief (1992, p. 146). She appeared, Stubbes insists, not sick but serene, prepared, joyous; above all, she projected a sense of control over her own image – a control that, of course, she quickly disavowed, remarking to those around her that her voice was God's voice, 'not the words of flesh and blood' (ibid., p. 146). The world in which Katherine suffered and died was a world in which childbirth posed real, ongoing dangers: given the likelihood of complications during the birthing process and the rudimentary aftercare available for her own body, Katherine might well have feared death for very good reason. And yet, despite the obvious and extraordinary vulnerability of her postnatal body, that body's suffering appears to have no tangible source in Katherine's story. Katherine's friends and neighbours gathered at her bedside not to witness the circumstances of her pain and eventual passing, but emphatically to praise her embrace of her body's destruction as a gift from heaven, proof of the power of the language of salvation, proof of God's power on earth in post-Reformation England. In this story, Katherine's dying female body is the flesh made Word.

Flesh made Word. When the female body in pain becomes the transcendent icon of grace; when the abused, violated, suffering woman takes upon herself the task of staging, Katherine Stubbes-like, a command performance of salvation: these are the moments on which this chapter dwells. Katherine's story, of course, is not exactly unique; the Protestant convention of the 'good death' circumscribed the ends of men and women alike in sixteenth- and early seventeenth-century England, turning their final suffering into the image of religious faith and thus into evidence of the enduring power of that faith to guide and support both local communities and the larger nation. And yet Katherine's story is also provocative precisely because it is not exactly *Katherine's* story: it is told posthumously by Phillip Stubbes, according to the dictates of his own religious and political aspirations. I ascribe agency – quite against the odds – to Katherine in my retelling of her deathbed scene only because Phillip does: his narrative deliberately casts his wife as stage manager of her blessed final hours in order to prove that her suffering is by her own faith, by her summons of God's word alone rendered grace. In turn, the tract in which he sets these events down becomes one of the period's pre-eminent conduct guides, transforming apparent description into pernicious prescription and painting a surprisingly tenacious image of the Godly woman as she for whom the painless peace that heralds death – regardless of its source; regardless of its potential preventability – is always the 'best gift' she 'can take' (Webster, 1997, 4.2.224).

I want to suggest that reading Stubbes's version of Katherine's final hours as something more than just Reformation business-as-usual may allow us to understand her deathbed spectacle as a specifically gendered event with a sinister underbelly. Below, I explore the troubling relationship between the language of painless, heroic suffering that encircles bodies like Katherine Stubbes's and the cultural staging of domestic violence against women in early modern England. I begin by examining the social and legal contours of 'reasonable correction' – the term by which the period's public discourse understood all manner of domestic battery. I pay particular attention to the ways in which household manuals written by Churchmen such as William Gouge, William Whately, and Robert Cleaver negotiate the slippage between their own absolute refusal to sanction the violent 'correction' of wives in the household and that violence's undeniable reality in practice. In the process of this negotiation these texts prescribe for battered wives the very kind of self-abnegating performance of salvation Stubbes scripts for Katherine. The vicissitudes of the good death and the promise of salvation it trails become handy shorthand for turning acts of domestic battery into public images of godly benevolence and English political fortitude.

Turning from legal and social history back to the stage, the rest of the chapter concentrates on how the slippery elision of violence and grace in the conduct literature translates to the larger space of Renaissance public representation. As I have been arguing throughout this book, early modern performance is an inherently dualist practice: it operates as one of the culture's central tools for the effacement of violence against women, but it also offers a space in which to interrogate the terms of that effacement – a space in which a history of violence's erasure may become visible while an ethics of its ongoing representation becomes possible. In light of this defining dualism, how might the battered woman's performance of salvation be transformed from a purely effacing practice into a querying, querulous in/visible act? The performance of salvation makes flesh into Word, makes a woman's body in pain lose its materiality en route to heaven's precious reward; the theatre, on the other hand, is a world in which words and flesh always cross-pollinate, materialize one another, and not always in stable or predictable fashion. On the stage word becomes flesh again, risking the integrity of salvation's performative apparatus and threatening yet another uncanny return.

Thomas Heywood's *A Woman Killed With Kindness* (1603), written and performed a decade after Phillip Stubbes's *A Crystal Glass for Christian Women* (1591), tells the story of Anne Frankford, young mistress of a petit bourgeois household who commits adultery with her husband's friend and familiar, Wendoll. Craving vengeance after discovering his wife and her lover *in flagrante*, Frankford exacts an unusual punishment: he stays his hand and banishes his wife to a house some miles away, but promises nevertheless to 'kill [her], even with kindness' – a promise she takes upon herself to fulfil (Heywood, 1985, 13.157). While Anne's actions and subsequent sentence may not offer exactly obvious parallels with the blameless life of Katherine Stubbes, her deathbed scene most certainly does (Gutierrez, 1994, p. 48). Like the Katherine of *A Crystal Glass*, Heywood's Anne produces a command performance of salvation according to the terms set out by her husband: she conceives, scripts, and executes her own death sentence, leaving him not only untouched by blame, but marked instead as her saviour and keeper of her sacred memory. She gathers her witnesses and demands that they authorize her undaunted faith and courage – which, of course, may only be done by turning a blind eye to the plain physical evidence of her suffering and of her husband's spectral violence-by-proxy. Unlike Katherine's, however, Anne's performance goes awry.

The punishment Frankford exacts and the death scene Anne subsequently performs – neither by any means typical, but both surprisingly revealing – may well have offered Heywood's audiences a rich site at which to explore the fault lines between the notion of household correction and the post-Reformation language of exemplary suffering circulating at the turn of the seventeenth century. After examining several key scenes in the play that worry these fault lines, I turn to British director Katie Mitchell's 1991 production of *Woman Killed* for the Royal Shakespeare Company. Mitchell's work on Heywood's play excavates the body that returns, uncomfortably, within the salvation narrative Anne Frankford performs, producing at every turn two bodies: the body the discourse of salvation expects *within* the body (the flesh) that discourse expels. Together, these contradictory bodies generate what Elin Diamond calls a 'double optic': the body expected and the body expelled '[bleed] through' one another in performance (2003, p. 11), operating as an uncanny duality that wrecks any attempt to read the violated female body as pure/clean/saved. Seen side-by-side, they interrupt household battery's easy disappearance into early modern spiritual discourse. *Heard* side-by-side, these bodies explode text, image, *and* expectation, bombarding spectators with body's lost voice, the cry of pain the Word so effectively suppresses.

Correction, salvation, and the companionate negotiation

> The soul is the prison of the body.
> (Foucault, 1995, p. 30)

Household corporal punishment in early modern England was a matter of what the law officially termed 'reasonable correction': wife-beating was simply good housekeeping. Emily Detmer notes that the '[p]hysical correction of wives was widely regarded as an appropriate means for ruling the household' and was the proper responsibility of the household head (1997, p. 275); 'reasonable correction' thus stands as a crucial example of the central differences between sixteenth- and seventeenth-century ideas of violence and our own (Amussen, 1994, p. 75; F. Dolan, 1996, p. 221). And yet, despite its apparent normalcy, 'reasonable correction' was a thorny problem for early modern legal and social commentators precisely because its limits were not clearly marked. 'As long as a husband did not kill, maim, or seriously endanger his subordinates or disturb his neighbours, he was not accountable for how he treated those in his care', Frances Dolan writes (1996, p. 218); Detmer adds, '[a]lthough the community got involved when it felt abuse was taking place, the culture did not enact laws that clearly distinguished between permissible and nonpermissible violence' (1997, p. 276; see also L. Davis, 1998, p. 64). William Heale makes the absence of clear legal language in this matter the subject of his extended treatise against wife-beating, *An Apologie for Women* (1609):

> [I]n the whole body of either Law, Cano[n] or Civil, I have not yet found (neither, as I thinke, hath any man els) set downe in these or equivalent termes, or otherwise past by any positive sentence or verdit *That it is lawfull for a husbande to beate his wife*.
>
> (1974, p. 28)

Heale uses this claim – that while the law does not set down any 'precise penalty' (ibid., p. 50) for wife-beating, it also never makes a positive statement of permission – to argue extensively for the moral, natural, legal, and spiritual *un*lawfulness of physical correction. On the other end of the spectrum, *The Lawes Resolutions of Womens Rights*, assessing 'correction' as a legal term, addresses the issue with surprising brevity. T.E. notes that 'The Baron may beate his *Wife*', but '[h]ow farre' such correction may have extended, exactly what kind of physical harm it may have permitted to be read as no harm at all, he (and the law) 'cannot tell' (1979, p. 128). The fact that *Lawes Resolutions* includes only one page on wife-beating in a 400-page text devoted expressly to women's legal issues, while Heale writes a 66-page treatise on that single problem alone, together reflect the extraordinary position the issue occupied in a culture only just beginning to come to terms with household battery as a genuine domestic crime.

Of course reasonable correction was not limitless. It could not threaten a wife's life without raising the spectre of plain, unjust violence, and thus of what 'reasonable correction' always already *could* be. Detmer argues that popular writings (such as Heale's) denouncing wife-beating were, by the turn of the seventeenth century, slowly succeeding in 'changing the cultural meaning' of household violence against wives (1997, p. 277), increasing the kinds and number of practices that might successfully be understood not as 'correction', but as what Laura Gowing calls 'extreme cruelty' (1996, p. 180). In practice, however, this shift in cultural meaning remained enmeshed in patriarchal optics – the culture's not particularly reliable ability to see the difference between an unremarkable beating and a life in danger. '[B]ecause the concerns of legal and religious institutions ran counter to the woman's own interest of escape from a dangerous situation', write Sara Mendelson and Patricia Crawford (1998, p. 143), '[e]ven when a wife was thought to be at risk, officials did not always act on her behalf. Male authorities often disagreed with women as to what constituted a dangerous degree of violence, or failed to take women's fears seriously' (ibid., p. 142). Given that household violence was embedded in early modern life as routine and not exceptional (Gowing, 1996, p. 184), wives suing their husbands for 'extreme cruelty' would have been at pains to describe their wounds precisely, and to muster as many neighbourhood witnesses as possible to speak with similar precision about physical evidence of harm (ibid., p. 209). Even then, neighbours could not always be counted on.[3] As Margaret Hunt argues, 'community intervention, even by women, should not be romanticized' (1992, p. 24) in cases of early modern spousal violence because few citizens, men or women, discounted the basic value of reasonable correction. Thus, women and the witnesses they could muster 'attended with as much care to the economic, mental, and verbal cruelty that gave violence its context' as they did to the actual physical proof of blows; this contextualizing allowed wives to cast their experience of violence as evidence of marital breakdown, a social problem with far greater cultural currency (Gowing, 1996, p. 210). For their part, men before the courts countered their wives' careful descriptions of wounds and bruises with descriptions of their exact actions and of the 'instruments' and parts of their bodies used in the corrective action, in an effort to paint that action as rational, reasonable, and, above all, as proof of a sense of care (ibid., p. 220).

In cases where a court determined that the line between correction and cruelty had been crossed, its only real recourse (other than sending husband and wife home again with a promise to live quietly together – often the preferred choice [L. Davis, 1998, p. 64]) was to institute a bed-and-board separation (Gowing, 1996, p. 181). As Heale writes, the argument that wives are legally bound to endure a beating from their husbands is specious, because the law permits a wife to 'depart from her husband' when her security cannot be guaranteed (1974, p. 46). Heale strategically overstates his case here: as Gowing's evidence suggests, the law would not permit a wife to leave except in thoroughly extreme circumstances for the simple reason that a bed-and-board separation dealt a blow to the household's integrity and, by extension, to the state's integrity. A life was preserved, to be sure, but at serious cost to patriarchal self-fashioning: the wife's bruised and bleeding departure scarred the image of a household smoothly run by a kindly patriarch whose firm (but gentle) hand guaranteed his status as natural ruler. These, then, were the stakes of permitting a woman's body in violence to appear (physically, legally, socially): no longer properly mediated by the discourses of household companionship, godly subservience, and private modesty that under normal circumstances shaped a wife's place in the home and the home's relationship to the patriarchal public sphere, that body became a cipher for something else – for what the discourse of patriarchal benevolence misses, what it buries in order to become.

Designed to assimilate all manner of husbandly brutality into its underlying logic of order-keeping, the loose rhetoric of 'reasonable correction' finally pre-empts the efficacy of much outside help, and reveals the extent to which early modern patriarchy relied for its very perpetuation upon a need to normalize the extremes of spousal abuse and family violence which were the 'inevitable corollary' of its routine use of force in the household (Amussen, 1995, p. 18). Because the early modern household was built on a gendered hierarchy which both permitted and expected a certain degree of physical correction, but which in turn insisted for the sake of its very legitimacy that such correction would always be delivered as kindly as possible, early modern patriarchy encountered its litmus test in the very extremes of violence for which it had no immediate justification. The question then became: how to transform those extremes of violence running counter to the core of patriarchal authority into the act and image of that authority, into the image of benevolence despite apparent evidence to the contrary?

In the absence of any broad cultural will to legislate the extremes of correction as unacceptable violence, early modern England turned to popular and widely disseminated sermons and conduct tracts to help shape its emerging understanding of the line between correction and violence against married women in the household (Detmer, 1997, p. 277). These texts managed the optical problem posed by the vague limits of 'reasonable correction' by condemning the practice of battery while *also* prescribing the godly wife's behaviour in response to a husband's beating. The tracts and sermons thus acted – and so I treat them here – as texts for social performance, scripts no less urgently actionable for a violated woman than was the instruction for rape victims in *Lawes Resolutions*, Bracton and Glanvill. With William Whately as one rare exception, religious (Protestant) conduct writers openly condemn husbands who are so vulgar, unkind, or un-Christian as to raise their hands to their wives,[4] but their opposition to the legal norm does not mean that they call for any kind of radical departure from the strict hierarchical ordering of early modern socio-political space. In fact, in its very refusal of the corrective norm the conduct material works to frame a moral continuity between the routine household violence that is (as it is well aware) an entrenched part of English Renaissance life and the extremes of violence which the doctrine of 'reasonable correction' can sometimes only with difficulty assimilate. The spectre of women's suffering ghosts 'reasonable correction' by embodying the contradictions (between benevolence and tyranny; between the 'natural' rights of male householders and their obvious, systemic abuse of those rights) that mark correction's limit. Conduct texts, in turn, work to erase the very appearance of contradiction within the discourse of 'reasonable correction' by offering battered women a framework through which they might stage their suffering not as injustice but as a form of cultural edification.

Nearly all late sixteenth and early seventeenth-century conduct writers counsel husbands against beating their wives, reminding them, as Robert Cleaver does, that proper rulers use their wives 'in all due benevolence' (1598, p. 119). Of a husband's duty, Cleaver writes: 'First, that he live with his wife discreetly, according unto knowledge. Secondly, that hee bee not bitter, fierce, and cruell unto her. Thirdly, that hee love, cherish, and nourish his wife, even as his owne bodie, and as Christ loved his Church' (ibid., p. 97). (Much later in his text Cleaver spells out the practicalities of this advice: a husband who 'seeketh to live in quiet with his wife must observe these three rules. Often to admonish: Seldome to reprove: And never to smite her' [ibid., p. 207].) This (entirely conventional) instruction generates two important effects. First, it ensures readers understand that, whatever may happen at the individual household level, and whatever the law may say, official Protestant patriarchy does not condone physical correction. Second, it implies that violence is a household reality but *should not be* – that is, it both acknowledges violence's reality ('hee bee not, bitter, fierce, and cruell') and yet simultaneously disavows its likelihood in a Christian household ('that hee love, cherish, and nourish his wife, even as his owne bodie, and as Christ loved his Church'). This latter gesture is crucial to the conduct material's handling of violence against wives, because even as it raises the spectre of such violence and, as Detmer quite fairly argues, attempts to reshape the existing relationship between correction and violence in post-Reformation England, it also places the material *effects* of 'extreme cruelty' *sous rature* in order to continue the regulatory fiction of cruelty's imminent disappearance in a properly Christian nation. Detmer focuses her reading of the conduct material on its instructions for husbands and their attempts to redesign the practice of wifely subordination around less outwardly visible forms of regulation and control (1997, p. 278), but I am more interested in what this material has to say to wives – and in what that advice may tell us about how the battered female body is made to sign when the distressing image of that body can no longer so easily be made to disappear.

In the religious conduct literature household violence moves from the somewhat fragile register of 'correction' (the legal register, where suits may not always be granted by court officials, but where they can always be argued by determined wives[5]) to the much more sure-footed register of Protestant doctrine and its official exegesis. In the portions of their texts devoted to women's duties to their husbands, many of the same conduct writers who admonish husbands not to beat their wives offer instructions to wives who have been subjected to the very violence they condemn. Here, again, the texts simultaneously acknowledge and deny that violence, reflecting it back to its female victims as a form of spiritual nurture proper to the holy estate of matrimony, the most important estate to which a Protestant man or woman could commit. As William Whately preached in 1624, 'Debarre marriage, and you bring the being of the World to a full point, yea, to a finall conclusion. Debarre marriage, and you shall have no families kept, no names maintained amongst men' (1976, p. 23). Thus, even at the wedding altar, the *Homily of the State of Matrimony* (1563) urged new brides to 'suffer an extreme husband' in order to win 'a great reward therefore' (1992, p. 20). Frances Dolan notes that while 'the homily attempts to redefine masculinity as nonviolent, it yet offers wives no recourse to beatings except prayer and patience' (1996, p. 172); I would add that this apparently paradoxical admonishment – husbands, be not violent; wives, suffer a violent husband patiently – is actually, for early modern religious officials caught between a rock and a hard place, a necessary elision. If stopping men from beating their wives was not especially easy, wives had to become the Church's first line of defence against the social and ideological fallout their beatings threatened.

In *A Godlie Forme of Householde Government*, Cleaver has this to say to wives about the obedience owed their husbands:

> [A]s it were a mo[n]strous matter, and the means to overthrow the person, that the body should, in refusing all subjection & obedience to the head, take upo[n] it to guide it selfe, and to commaund the head: so were it for the wife to rebell against the husband. Let her then beware of disordering and perverting the course, which God in his wisedome hath established: and with all let her understand, that going about it, she riseth not so much against her husband, as against GOD: and that it is her good, and honor, to obey God in her subjection & obedience to her husband. If in the practise of this dutie, she finde *any difficultie or trouble, through the inconsiderate course of her husband,* or otherwise; [...] *let her humble her selfe in the sight of God,* and be well assured that her subjection and obedience is acceptable unto him. Likewise, that *the more that the image of God is restored in her, and her husband, through the regeneration of the holie Ghost, the lesse difficultie shall she finde in that subjection and obedience* [...].
>
> (1598, pp. 227–8, my emphasis)

The husband's blows (already spectral here in Cleaver's cautious language) recede as the wife's performance of spiritual obedience fills out the frame. It does not matter that he hits her so much as it matters that *she does not hit back*: as she models Christian patience and fortitude in the face of 'difficultie', she sutures the rips in the household fabric her husband's cruelty makes. In her exemplary passive resistance the abused woman becomes a cultural icon, a mirror, like Katherine Stubbes, of proper Christian womanliness. She stands for the endurance of her Christian marriage and the continuation of the Reform state – and threatens the dissolution of neither. *Of Domesticall Duties*, William Gouge's enormous and comprehensive 1622 tract, offers very similar advice to wives subject to what he calls 'reproofe' (1976, pp. 320–1). A husband's duty is to 'rebuke' his wife; as she owes him obedience, she must look upon such a rebuke as an opportunity to prove her subordination in a moment of that subordination's explicit trial. The 'rebuke' is thus a spiritual opportunity, even a test, for the wife. And when it is unjust or 'bitter' (ibid., p. 320), it is ever more a test: 'it is a grace, a glory to her: a matter that deserveth praise and commendation' in particular because it is *'acceptable to God'* (ibid., p. 321).

Whately offers the most explicit articulation of this rhetorical strategy in his popular wedding sermon, *A Bride-Bush, Or, A Direction for Married Persons*, first published in 1617. Reminding his audience that 'patient suffering is one part of subjection to authoritie' (1623, p. 210), he divides the physical harm to which a wife may be subject into two categories: that which she deserves as a result of misbehaviour, and that which proceeds unjustly from a tyrannical husband. In the former case, Whately argues that the wife 'must thanke herselfe' for the beatings (ibid., p. 211), 'and not alone be patient under them, but carefull also to receive them fruitfully, making conscience to reforme the faults that have procured them' (ibid., p. 212). In the latter case, he implies – as does Gouge (1976, p. 320) – that the wife must thank the Lord for providing her with a trial of her patience and fortitude, and take the opportunity to show herself truly worthy of grace: '[M]eeknesse and quietnesse is not worthy the name of meekenesse and quietnesse, which can onely then bee quiet, when no cause is offered unto it of unquietnesse, by hard and unjust usage' (Whately, 1623, p. 212). Finally, like any good Protestant martyr, the wife ascends to the role of religious exemplar: her 'chast behaviour united with feare' offers her husband a spiritual model to emulate, thus guaranteeing them both divine sanction (ibid., p. 215). In all cases, the reward that these writers insist awaits a brutalized wife can only be accessed by the wife's absolute capitulation to her husband; she might seek out her neighbours or the magistrate in extreme circumstances, but she must always return home (ibid., pp. 214–15).

Labelling it variously as a moral lesson, as a gift of grace, and even as female agency, writers such as Cleaver, Gouge, and Whately recuperate extreme cruelty as the very image of a benevolent God, a God reassuringly at work in the world despite the manifold perversions of His will enacted by the wayward husbands assigned to stand in His place. As I build this argument I would like to be clear that I am not suggesting that post-Reformation Christianity is somehow inherently to blame for the continuing struggle of early modern wives against domestic abuse, even though Reformation doctrine *is* inherently patriarchal and therefore tends, like all patriarchal structures, wittingly or unwittingly toward misogyny and its material effects. I am a secular feminist scholar, but I nevertheless believe that we must make space for a nuanced appreciation of all religious doctrine, when our research demands its discussion, in order to account for the manifold critical possibilities – including feminist possibilities – such complex forms of discourse embed. My concern in this section lies specifically in how post-Reformation religious officials negotiated the gaps between what they perceived to be an essentially benevolent doctrine and its less-than-benevolent practice, and in how they enforced the preservation of a strong patriarchal order amidst the problems created by household patriarchs who did not always appreciate the challenges and responsibilities of their roles in the game.

The notion that a battered woman might be authorized, indeed required, to perform not only the promise of her own salvation but also that of her abusing husband needs to be understood within the framework of companionism, early modern England's dominant marital model and the one to whose detailed explication conduct writers devote themselves. Companionism is based on 'mutual subjection', the notion that husband and wife are one another's spiritual supports, and that each owes duties of care to the other that extend, importantly, to *'mutuall care for one another's salvation'* (Gouge, 1976, p. 238).[6] 'Mutual subjection' does not, importantly, advocate spousal equality; the term implies a hierarchy in which 'subjection' is the controlling idea. Husband and wife are spiritual partners but not social equals: they are understood to be one flesh – but one in *him*. Christine Peters argues that early modern men and women would have accepted without question the legitimacy of this seemingly imbalanced construct, understanding that 'mutual subjection involved distinguishing between two types of subjection: subjection of reverence and subjection of service' (2003, p. 315). A wife owed her husband both reverence (as her master) and service (as her fellow Christian), while a husband owed his wife only subjection of service, the 'common' Christian duty that all believers owed one another. Companionism's major purpose was thus not to advance women's rights in the home but to ensure the proper ordering of both the home and the larger Christian public sphere (ibid., p. 315). However, because it embedded a measure of spiritual equity within a strict and rigid social hierarchy, companionism's complex structure also required plenty of discursive and performative labour to maintain.

Companionism has been controversial for contemporary feminist scholars. Many have argued that the very notion of 'mutual subjection' contains inherent contradictions that would have been visible to the early moderns and would have promoted, among other things, insubordination and increased household violence (Davies, 1981, p. 63; McLuskie, 1989, p. 48; Lucas, 1990, p. 230; Gowing, 1996, pp. 223–7; Comensoli, 1996, p. 11). Others insist that the 'contradiction' argument relies too heavily on a trans-historical feminist perspective and fails to account for the very different conditions of gendered understanding at work among the early moderns (Wayne, 1992, pp. 14–16; Peters,

2003, pp. 314–16). Rather than argue that companionate logic is simply accepted, wholesale, at this moment in history (evidence of normalcy and evidence of acceptance are two very different things), or that it is simply unjust and unacceptable in any historical moment, I want to suggest that this logic, like the body of conduct literature that articulates its terms and maintains its authority, exists as a fragile, continuous negotiation between what is and what could be, between what it says and what it disavows. Reading from a historicist position informed by contemporary feminist understandings of domestic violence, we might note that the very idea of benevolent companionism and its 'fiction of subsumption, of two becoming one' implies a certain level of violence against wives (Dolan, 1994a, p. 99; see also Detmer, 1997, p. 293). While conduct writers frequently used the companionate commonplace of husband and wife as one flesh to support their arguments *against* wife-beating (Gouge, 1976, p. 391, for example), to many husbands 'obedience, not unity, was the keynote of their understanding' of the marital model that attempted to yoke these ideas seamlessly together, and 'violence of varying degrees, threatened and real, was the tool with which obedience was to be enforced' (Gowing, 1996, p. 223). In other words, while conduct writers were clearly troubled by the gap between doctrinal understanding and religious practice in the household, the unstable relationship between *conduct theory* and its practice – for both husbands and wives – was also a serious problem to which we can assume they were attuned. In *Of Domesticall Duties*, Gouge makes pre-eminent the spiritual partnership that obtains between husbands and wives, and as a result has to backtrack in order to control that assertion's potentially volatile implications. In his second and third treatises he spends some time considering the areas over which husbands and wives ought to have joint jurisdiction (including the governance of servants and children, as well as in matters of the soul), but then insists that one cannot infer general equality from these singular instances, for '[t]hough there *seeme to be* never so little disparitie, yet God having so expressly appointed subjection, it ought to be acknowledged: and though husband and wife may mutually serve one another through love: yet the Apostle suffereth not a woman to rule over the man' (1976, p. 272, my emphasis). The structure of this rhetoric, coupled with the sheer number of times Gouge reiterates the importance of wifely subjection throughout his text, reveal the extent to which tracts like his based their messages on the often unwieldy conflation of a potentially progressive mutualism with old-style patriarchal dominance and submission.

Gowing's comment on the subtle interrelationship between unity, obedience, and violence within companionate doctrine, read alongside Gouge's latent anxieties about the uncontrollable consequences of some of his own advice, suggests something important: not just that companionism embeds the potential for violence, but also that violence against wives may in fact lie, unacknowledged, at the base of companionate doctrine. In her long chapter on 'the heroics of marriage' in *The Expense of Spirit*, Mary Beth Rose notes that post-Reformation English family life is a set of private relationships enacted on a very public stage (1988, p. 119), and that early modern marriage is, similarly, a private-public arena 'in which the individual can struggle and meet death or defeat, triumph or salvation' (p. 121). The weapons with which husband and wife fight are (at least, in theory) not arms but 'inner strength': they muster the courage 'to suffer and endure' (ibid., p. 122). Rose's 'heroism of endurance' (p. 123) provocatively glosses my reading of household violence as something that the conduct literature is at pains to stage-manage. Via her argument, we might consider how the *idea* of household violence, even in its most extreme forms, is central to shaping private, domestic companionism as a form of public Protestant heroism – just as the idea of martyrdom was central to the construction of the Reformation Protestant subject. (We might also note that companionism preceded the Reformation [Davies, 1981], though the latter, with its focus on marriage as the ideal Christian estate, gave it serious traction. Like all pre-Reformation practices, companionism undoubtedly came with baggage the new religious order needed to assimilate.) To understand – as Whately and his fellows do – the battered wife as a household exemplar within a marital structure based on the promise of heroic suffering is to understand 'extreme cruelty' as more than just an undesirable consequence of changing cultural and religious attitudes toward women and corporal punishment. Rather, it is to understand the *possibility* for such cruelty as undesirable in practice, but at the same time paradoxically central *in theory* to early modern English constructions of a heroic, Christian union. This is not to say the early moderns secretly wished to destroy their wives; it is to say, however, that the threat of violence, however outwardly condemned, may well have been in every way socially and spiritually sustaining for the early modern household. The early modern woman's battered body thus becomes the perfect paradox: both the threat of companionism's radical failure *and* the *sine qua non* of mutual subjection, an essential stage on which to declaim the marriage that saves and endures.

Making a spectacle of salvation

> 'Twas I that killed her.
> O, the more angel she,
> And you the blacker devil!
> (*Othello*, 5.2.131–2)

This same paradox moves Thomas Heywood's *A Woman Killed With Kindness*. The play has been called a dramatized conduct book (Bromley, 1986; Panek, 1994; Creaser, 2005), and not unreasonably given that it quite neatly reimagines Katherine Stubbes's exemplary Christian death. Frankford banishes his wife, Anne, from his household after he discovers her in an act of adultery with his friend and familiar Wendoll. Chastened and eager to prove her penitence, Anne then orchestrates her own death by fasting; the fast culminates in a deathbed spectacle of redemption and salvation unparalleled in the drama of the period. Nearly all of Anne's witnesses praise both her Christian fortitude and Frankford's successful spiritual care of her in choosing banishment over other, more plainly vicious forms of punishment; their marriage bond is restored and his status as benevolent household head is confirmed even as Anne passes away. And yet *Woman Killed* is not a simple rehearsal of the Stubbes scene, nor a simple dramatization of the advice to husbands and wives offered by writers such as Whately, Cleaver, and Gouge. At its most sanctified moments, the play is subtly but provocatively critical of the motives and assumptions underlying the conduct literature with which it circulates as fellow traveller in the Jacobean public sphere.

The status of the 'kindness' in *A Woman Killed With Kindness* has long been a source of debate for scholars trying to determine whether the play's endorsement of Frankford's behaviour is earnest or ironic. Some critics read Frankford straight, as a model husband whose refusal to resort to violence dramatizes perfectly the conduct literature's apparent rejection of physical correction in favour of gentler means of domination and submission (Wentworth, 1990, p. 157; see also Kiefer, 1986, p. 91; Rossini, 1998, p. 114). Others have suggested that when Frankford banishes Anne from his home, he diligently places the maintenance of proper household order and the integrity of his family's honour above all else, appearing once again as an icon of good husbandry (Bromley, 1986, pp. 266–7; Orlin, 1994, p. 151; see also Creaser, 2005). On the other side of the coin, Paula McQuade (2000), Lyn Bennett (2000), Manuela Rossini (1998), Nancy Gutierrez (1989 and1994), and Jennifer Panek (1994) have argued to sometimes different ends that the play is deeply critical of early modern family structures and marital ideals, and that, in Frankford, Heywood offers us a model for how *not* to govern a household, how *not* to handle a wife. For these writers, the play exposes the violence at the very heart of 'the liberal-humanist marriage' in which the viciousness of a husband's absolute rule is 'masked as kindness' (Rossini, 1998, p. 116; see also Panek, 1994, p. 370). And yet in *Woman Killed* murder is not simply 'masked' as kindness; it is given no status *other than* kindness – a subtle but crucial distinction. Frankford technically does no more than speak; Anne's death appears as cleansing fast and sincere repentance, and in its wake Frankford's promise to 'kill' her with kindness materializes as redemption and salvation. *Woman Killed* does not simply reveal the often palpable, but broadly disavowed component of violence within patriarchal absolutism and mutual subjection; it explicitly engages and finally 'outs' the recodification of violence on which the propagation of that absolutism and subjection rests. This play is less about the politics of forgiveness or the violence of the patriarchal everyday than it is about the investment the patriarchal everyday makes in the simultaneous act of violence against a wife and its legal and social eradication.

'Kindness' is less a mask than a Janus (Orlin, 1994, p. 261) for those early moderns committed to the ideals espoused in conduct writings, and the play's relationship to the term is, for me, not either/or but both/ and, neither a hopeless contradiction nor a stable signifier through which to channel Frankford's 'true' nature, but a tricky double optic through which to understand his profoundly complicated, and complicating, uptake of conduct logic. During his discussion in *A Bride-Bush* of whether or not a husband may beat his wife, Whately employs 'kindness' in two related but distinct ways. He argues first that husbands must find alternatives to physical correction whenever possible; among those alternatives may be 'with-drawing from [one's wife] the plentifull demonstrations of kindnesse, and fruits of his liberalitie, and by abridging her of her libertie, and the injoyment of many things delightfull' (Whately, 1623, p. 106). This advice is reminiscent of Petruchio's strategy for handling Katherine in *The Taming of the Shrew*, and on first glance it also seems very much the sort of 'kindness' Frankford shows Anne when he banishes her. But it is not the only sort. A short time later, Whately considers what husbands may do if their wives fail to respond to patient, non-corporal attempts at correction:

> but even blowes, after patient forbearance, after much waiting for amendment without blowes, and so applied, that it is apparant, a

man seekes not to ease his stomacke, but to heale his owne flesh with a corasive [corrosive], when nothing else will doe it (even blowes I say with these limitations) *may well stand with the dearest kindnesses of matrimony.*

(1623, p. 108, my emphasis)[7]

Whately is extremely cautious here to distinguish between moderate, necessary correction as a last resort and violence that may be construed as cruelty ('a man seekes not to ease his stomacke'); nevertheless, he links kindness unequivocally with 'blowes' in this passage, giving his earlier use of the term a fresh valence. Kindness is turning the other cheek, but it is *also* a switch to the back or a slap to the face. Kindness is both blows and the refusal to resort to blows. Kindness is a husband's last resort in his attempt to assert his position of dominance over his wife in their joint household, and it is a wife's duty to take the measure of her husband's beating *as* a kindness, as I noted in my earlier reading of Whately's advice to wives, and to 'thank him' for her pains.[8]

Whately's 'kindness' lets us read Heywood's title through a new lens.[9] Perhaps the problem is not that Frankford is *not* kind, and perhaps the problem is not that he is *too* kind: perhaps the problem is that he is just kind enough, that he understands 'kindness' literally, in Whately's doubled sense of the term. Frankford turns the other cheek when he banishes Anne from his sight, but he *also* couches that banishment in the language of deliberate physical and psychic cruelty: he will 'torment' her 'soul'; he will 'kill' her with 'kindness' (Heywood, 1985, 13.156–7). Frankford uses 'kindness' simultaneously to disguise and to expose his intentions, but above all to claim for his violence the status of (sinister, inverted) spiritual nourishment. Frankford isn't a failed householder, and he isn't its ideal: he is its dark underside, the uncanny figure hidden within a popular body of social literature that will refuse to sanction violence against women even while allowing it re-entry through the back door. In killing his wife and *calling* it kindness, Frankford exposes the tyranny implicit in the conduct canon's promise to wives that being beaten up amounts to a heavenly prize: her own and the household's salvation.

My reading of Frankford's double-kindness pivots upon several key moments in scene 13. Frankford and his servant Nick have returned from a false journey designed to take them from home and leave Anne surreptitiously to her lover. Before he enters the bedroom in which he expects to catch his wife and friend in the act, Frankford prays for 'patience to digest my grief / That I may keep this white and virgin hand /

From any violent outrage or red murder' (13.31–3). This prayer for patience is conduct pitch-perfect, and should we doubt it, Frankford's gentlemanly prayer contrasts with Nick's *im*patient aside once his master leaves: 'Here's a circumstance! / A man may be made cuckold in the time / That he's about it' (ibid., ll. 35–6). Frankford returns, having discovered the lovers but not woken them. His second prayer shifts accordingly:

> But that I would not damn two precious souls
> Bought with my Saviour's blood, and send them laden
> With all their scarlet sins upon their backs
> Unto a fearful judgement, their two lives
> Had met upon my rapier.
>
> (ll. 45–9)

Once more, Frankford plays the role of the proper household patriarch. Recognizing his Christian duty of care to both Anne (as his wife) and Wendoll (as his friend and guest), Frankford frames his refusal to strike in the language of spiritual protection. As he prepares to go 'in to wake them' (l. 66) he calls once more for God to grant him 'patience' (l. 65), but now, clearly, that call is not enough. Moments later Wendoll flies across the stage in his nightgown as Frankford chases him with his sword drawn. Unable to contain his own rage, Frankford finds his hand stayed by the maid, whom he thanks and calls 'angel' for protecting him 'from a bloody sacrifice' (ll. 69–70). 'Patience' operates in this brief episode as 'kindness' will moments later: it is a performative utterance Frankford deploys in order to stabilize his gentlemanliness and his godliness. And yet as the scene progresses, his call for 'patience' also appears as a convenient means for Frankford to deflect attention (his own and ours) from his obvious and overwhelming urge to bloody violence. Even as he prepares to attack Wendoll, Frankford maintains that he is the kind of man who turns the other cheek. Finally, he does resist his urge, banishing Wendoll to contemplate his wrong against his host's 'many courtesies' (l. 73). For Frankford, then, 'patience' lies at the heart of a sincere prayer for godly strength, but it is *also* the claim to goodness one makes after murder has been narrowly avoided.

With Wendoll gone, Frankford must deal with Anne. She is shamefaced and begs for punishment (13.91–106; ll. 133–45), but on one condition: she prays that Frankford send her 'Perfect and undeformed to [her] tomb' (l. 101). Anne craves a martyr's death; she wishes her body to remain whole so that she may be seen, eulogized and remembered as a cautionary example for the women who come after her (ll. 144–5). But she also, apparently, craves protection for her husband, for if she dies 'undeformed' no one may say that he abused her, or that he acted toward her body and her soul in anything but the best spirit of 'correction'. Anne – a wife both ideal and unreal in her perfect self-subjection (Barker, 2007, p. 170) – here enacts the conduct script to the letter as she plants the seeds of Frankford's course of action; for his part Frankford both heeds and resists the import of her prayer as he finally passes sentence on Anne. He settles on banishment from bed and board (ll. 158–72), an entirely appropriate punishment under the circumstances, but he also shapes his sentence as a direct reply to Anne's earlier call for a martyr's death:

> My words are registered in heaven already;
> With patience hear me. I'll not martyr thee,
> Nor mark thee for a strumpet, but with usage
> Of more humility torment thy soul,
> And kill thee, even with kindness.
>
> (ll. 153–7)

In this short passage the play's competing articulations of 'kindness' collide with a force that startles, provokes a double-take, demands that audiences hear Frankford's words both ways – for the kindness they promise as well as for the cruelty that kindness sustains. These words style destruction, openly, as preservation ('with usage / Of more humility *torment* thy soul') and threat as care ('kill thee, even with kindness') as Frankford uses the pose of household spiritual guide to disguise, indeed to shape, his latent vengeful urges. His sentence turns the language of blessing and grace *itself violent*; it imagines not a clear path to redemption but radiates a disquieting ambiguity. Rather than 'bless' her with blows, Frankford's voice sears, brutalizes, promises Anne in her exile the 'loss of [...] family, health, love, honor' that is 'the classic substance' not of a blessing but 'of a curse' (Gross, 2001, p. 164). He answers the trauma of his own loss with the promise of parallel losses for Anne (ibid., p. 165), confirming how vulnerable and traumatized her actions have left them both (Barker, 2007, p. 188). Sending his wife away, Frankford invokes companionism's doubled body directly as he reminds Anne that her act of betrayal against their marriage has left them, literally, rent. 'It was thy hand', he says, 'cut two hearts out of one' (13.186). Lest there be any doubt in anyone's mind: she, not he, is the author of the violence that moves this scene.

Frankford's curse makes trouble for Anne as she works to bring the play to its conclusion. Anne is the play's penitent whore, but she is also the wife who aspires to household heroism – and it is thus on her performance of penitence and salvation that this domestic morality tale depends. Anne is eager to play these roles: she welcomes death (13.133); she 'would have this hand cut off, these my breasts seared, / Be racked, strappadoed, put to any torment' to 'whip but this scandal out' and redeem her honour and that of her husband and children (ll. 136–8). As I have suggested, however, *Woman Killed* is no typical morality tale: it has a de-realizing relationship to the social literature it appears, at first glance, to take as its performative base. Frankford's curse is troublesome because it produces neither violence nor the gentlemanly refusal of violence (both, in their different ways, sanctioned by the conduct canon), but something queerly in between.

William Whately was an extreme voice in early modern England, but while certain of his views on wife-beating were controversial his prescriptions for domestic violence's best outcome were entirely standard. What a husband metes out 'kindly' (or unkindly, for that matter) a wife must accept 'meekly', in both Whately and Gouge's preferred term: at the moment of the blow, however justly or unjustly dealt, a wife's job is to take her husband's violence as a trial from God. Frankford stretches one possible interpretation of this prescription to its extreme. In an uncanny mimicry of the conduct literature's premise that a husband's violence is ultimately a wife's to have and to hold, Frankford's promise to kill Anne with kindness produces violence that *has yet to be* – violence that Anne literally must bring into being before she can transform it into an image of his, and God's, benevolence. Frankford's curse is violence for which Anne must take both material *and* performative responsibility. A curse is a performative act: it is a declaration of future violence, and as such it requires both the belief and the complicity of its target in order to produce the pain and dispossession it promises. For Anne to play the role of the penitent, she needs something to play against, something to recuperate as evidence of her (and her household's) spiritual regeneration. Frankford's curse alone cannot provide her with performance material: it is '[a] mild sentence' (13.172), she declares blandly. Clearly, blows would have done better: in the face of blows, Anne would have at least known what to do, what to say, how to behave. Now, in order to make her hoped-for exemplary death scene possible, Anne must use Frankford's curse to become the active agent of her own suffering – a suffering she will *then* recuperate as blessing, having masked its source, in her deathbed performance of salvation. Thus, at the same time the curse demands Anne materialize the violence it promises it *also* requires her to resuscitate Frankford's wayward speech, cloak its exposed cruelties, resolve its ambiguities. Her task is not to overwrite her suffering with the language of goodness – a doctrinally standard, even expected, gesture – but to *generate* the image of her suffering as a by-product of the language of goodness and then to disguise the relationship between the two. To get to her big scene, Anne must fight Frankford's complex utterance on its own terms, engage in a tricky exegesis, and negate half its truth-value (the half that is more 'kill' than kind) with competing testimony. She must stage a performance of salvation in which the object of attempted recuperation is the language of 'kindness' itself.

To produce the suffering Frankford's curse demands Anne chooses to fast. In the early seventeenth century fasting was still an accepted purification ritual designed to prepare women 'for being good wives and mothers' (Frey and Lieblein, 2004, pp. 46, 50); in the context of Frankford's curse, however, Anne's fast is also a striking incarnation of oral violence against her own body. Christopher Frey and Leanore Lieblein argue that, while fasting retained some of its spiritual value at the time of the play's first production, through the seventeenth century the 'starved female body' came to be read as pathological as well as religious, more 'sickly' than 'saintly', as it morphed into an object of clinical scrutiny (ibid., p. 47).[10] Frey and Lieblein position Anne at the crossroads of three competing historical narratives of female fasting – the saintly ascetic, the self-sacrificing Renaissance mother, and 'female starvation as illness' – which they label 'contradictory' (ibid., p. 47). Once more, however, I propose that what appears at first as contradiction may profitably be understood as both/and, a critical doubling. When Anne takes Frankford's curse in at her mouth, she inadvertently marries the fast's promise of redemption through ritual cleansing and spiritual sacrifice to the emerging 'problem' of the self-starved female body violated by wasting. Her hoped-for kindness becomes a kill after all: she enacts (rather than resolves) the troublesome double optic produced by Frankford's queer promise, marks her body with the radical confluence of cruelty and kindness embedded in his words and in the larger domestic and spiritual discourses from which they take their cue. Anne's open mouth cannot simply testify to Frankford's goodness because it is *also* the site of physical devastation (the body wasted from lack of food) and spiritual invasion (his curse literally internalized).

Under these vexed circumstances, Anne's by-the-book performance of salvation mocks its own intentions. Frey and Lieblein argue that Anne's fast returns her body to the stage at the eleventh hour, answering its near-total effacement earlier in the play and in Frankford's 'mild' sentence (2004, pp. 60, 61). While I disagree with their argument that Anne responds to Frankford's 'passive inscriptions' by 'substitut[ing] the *active choice* to mutilate her own body' (ibid., p. 59, my emphasis), I support their reading of the final scene as an uncanny return. But what exactly has gone missing? Not just Anne's body: the source and meaning of the violence committed through and against that body have similarly been effaced. Anne's uncanny salvation scene brings all these missing back to the stage. Propped up in her bed at centre stage, Anne assembles witnesses to read the signs of 'fault writ in [her] cheek' (17.56), but Sir Charles's reply is ambivalent at best: 'Sickness hath not left you / Blood in your face enough to make you blush' (ll. 58–9). Anne hopes her body will record her penance and thus the 'kindness' of the fast; she must confront instead the consequences of its readily visible physical and psychological effects. Hers fails as an ordeal body because it records only its own disintegration; no longer fit to sign beyond itself it now signs only its (damaged) self, the damage 'kindness' has wrought. After Anne's passing, Sir Francis tells Frankford: 'Brother, had you with threats and usage bad / Punished her sin, the grief of her offence / Had not with such true sorrow touched her heart' (17.133–5), but in the wake of Anne's wasted body the comment sticks hard. What, really, constitutes the difference between the blows Frankford *could* have given Anne and the blows she gave herself? Only language, once more – the language of forgiveness Frankford finally offers and Anne gratefully receives, as true blessing this time (ll. 114–20). But language is also not given a free pass here. When the assembled company pledge, along with Frankford, to 'die' with Anne in their grief, Nick responds to the contrary: 'So will not I! I'll sigh and sob, but, by my faith, not die' (ll. 99–100). Nick, unlike the other bedside witnesses, says what is true, not what is designed to mask the truth in order to make everyone feel better. Anne is going to die, and they are going to cry, and then it will be over. Lest there be any doubt in anyone's mind, Nick's aside very clearly reminds us: Anne, and nobody else, is the victim of this scene. Our job, as her witnesses (on stage and off), is to recognize as much: to resist Frankford's platitude and hear instead the receding voice, ventriloquized here by Nick, of the buried body in pain.

Body doubles, Babel's voices

In Katie Mitchell's 1991 staging of *A Woman Killed With Kindness* for the Royal Shakespeare Company,[11] that receding, resisting voice overtakes centre stage during Anne's deathbed scene. The lights, as they have been throughout the performance, are almost painfully low. Anne, played with intense, fervent physicality by Saskia Reeves, lies in a small, cradle-like bed lit by a single, onstage spot; the spot focuses what little light this scene affords squarely on Reeves's ravaged body, which spectators must strain to see through the darkness of the scene and past the bed's high side-boards.[12] If we wish to be among Anne's witnesses, clearly we must do more than simply follow her drama of redemption and salvation by rote. The stage set-up asks for our labour, our concentration, our physical as well as intellectual effort on Anne's behalf. As Reeves's Anne calls out to her onstage witnesses her head lolls backward, too weak to hold itself up; as it rolls, it gestures toward the oversized, rustic wooden cross that dominates Mitchell's playing area and forms an oppressive headboard for the comparatively tiny bed. For those sitting in some of the best seats in the house, directly in front of centre stage, the cross mediates the view; Reeves can barely be seen for this heavy, ominous remnant of Christ's own suffering and salvation.

Frankford arrives, played by Michael Maloney with similar manic energy. Reeves gasps and tries to raise herself but her body, shockingly wasted, can barely manage to sit up. Anne's 'zeal to heaven, whither [she is] now bound' (17.82), threaded through a voice raspy and ravaged by extreme physical suffering, becomes no happy testimonial but a Brechtian *verfremdung* akin to Sonia Ritter's moments of collapse in Deborah Warner's then-recent *Titus Andronicus*. Reeves's voice can speak only the weight of the corporeal unspoken in Anne's words, a far cry from crying heaven's, and Frankford's, praises (I hear Reeves's Anna and think of Kate Stubbes: what must her final testimony have sounded like?). Maloney, meanwhile, weeps openly and with almost disturbing visceral commitment as he kneels at Anne's bed to hear her cries for pardon. In place of his earlier pledge to kill her with kindness, Frankford now proffers his 'wish to die with' Anne (l. 97) through convulsive tears that stick gutturally in Maloney's throat. Anne's assembled witnesses, not surprisingly, are not convinced: they deliver their prescribed response, 'So do we all' (l. 98), flatly, distantly, openly refusing the obscuring sentiment of Frankford's second promise even as they offer the testimony this performance requires. Nick's aside is spoken from the shadows in similarly emotionless tones; his words collide with Frankford's over-determined prayer in the air above Anne's dimly illuminated body, refracting the play's many conceits of household duty and religious devotion as hollow, violent. Oblivious to this effect, Anne joyfully receives her husband's pardon and his restoration of her title of wife and mother; she remains submissive before both him and the promise of heaven's freedoms (l. 121) even as her body shudders in pain (Barker, 2007, p. 187). The moment risks sentiment but for Maloney and Reeves's virtuoso physicality: their tears and gasps, and Reeves's extraordinary physical struggles, appear simply too real for seamless, digestible reception (Figure 2). As Maloney closes the play with Frankford's final promise to erect a golden epitaph in Anne's name, his overwhelming grief gives way, on his final line, to an eerie intonation: 'Here lies she whom her husband's kindness killed' (l. 140). He directs the line to the audience, over-enunciating the 'd' in 'kindness' and giving the last syllable a sibilant vocal echo that seems to make the word unravel. This is no 'good death' scene, and Anne's witnesses both on and off stage are understandably at sea, unsure how to behave, to perform, to receive her queer spectacle of salvation.

In her comprehensive reading of this production, Roberta Barker explores what she characterizes as Mitchell's radical naturalism, an intensely realist production style complicated by a lack of historical fixity or consistent psychological characterization. Mitchell's naturalism is informed by early realism's commitment to the actor's physiological

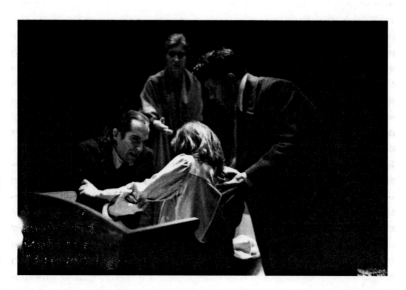

Figure 2 Michael Maloney as Frankford and Saskia Reeves as Anne in Katie Mitchell's 1992 production of *A Woman Killed With Kindness* for the RSC. Photo by Leah Gordon, with her kind permission.

response systems, and she has spoken at length in interviews about a working process she characterizes as a Stanislavskian technique (Giannachi and Luckhurst, 1999, p. 97) based at least in part on the method of physical actions and telescoped through her study with, and observation of, directors such as Anatoly Vasilyev and Lev Dodin in Russia, Poland, Lithuania, and Georgia. Mitchell's process involves lengthy, intensive research and textual preparation (Shevtsova, 2006, pp. 7–8; Manfull, 1997, pp. 103–4) as well as plenty of work on text and intention, cause and effect, with her actors in the rehearsal room. Actors work both 'outside in and inside out' (Shevtsova, 2006, p. 10), using specific, repeated physical actions in a manner designed to produce particular 'targeted' emotions within the audience (pp. 10–11). In a 2006 interview with Maria Shevtsova, Mitchell positions her approach to naturalism specifically against post-war British realism and its focus on well-spoken text over expressive, emotionally authentic bodies. 'Much mainstream theatre here is very preoccupied with words and hearing them spoken clearly', she argues; '[t]here is less interest in representing human behaviour accurately [...]. Expressions of human behaviour in theatre tend to be either exaggerated or too discreet or made up of self-conscious and artificial gestures and sounds. This type of theatre does not interest me' (ibid., pp. 8–9).

Thinking Mitchell's techniques through her *Woman Killed*, Barker notes that the production's 'juxtaposition of elements' (including costumes mixing early modern, nineteenth- and twentieth-century styles, a set design reminiscent of American frontier aesthetics, and strong Yorkshire accents that in no way matched the Georgian Catholicism of the characters' religious practice), alongside Mitchell's refusal to produce 'consistent journey[s] in the performances' of her actors/characters, made reviewers profoundly uncomfortable. These production choices disrupted not only normative audience expectations of a night at the RSC, but also expectations of what naturalism – a theatrical style that typically promises to reveal the hidden truths of its characters' psychic lives – should look and sound like (Barker, 2007, p. 167). For Barker, Mitchell's 'simultaneous engagement with and disruption of naturalistic modes' allowed her to 'represent a society in which power relations were not fixed and irreversible, but mutable and fluid' (ibid., p. 168). Her critical realist style – not far from Brecht in its effects, but in no way Brechtian in process – created a world sustained, visibly and audibly, by regulatory fictions haunted by practised disavowals.

I want here to extend Barker's nuanced reading of how Mitchell's radical naturalism signified in this production, particularly on the bodies of her actors and particularly in moments of violence and suffering. As I hope my evocation of the pivotal final scene demonstrates, Mitchell's performers disrupted naturalism's acting conventions by producing not just too little realism, but also *too much*. Reeves and Maloney troubled what we may imagine to have been more than a few seasoned spectators' conventional 'realist' expectations by pushing their characters' commitments to normative behaviours and beliefs so far that the performing body appeared, literally, to rebel. Pressing psychological realism to its critical and ethical limit, Mitchell's actors gasped and sobbed to the point of retching in a manner I found very hard to watch; they staged a collision between the bodily conformities expected by conventional naturalist mimesis and the affective and physical disavowals that guarantee the naturalist body's adherence to a rigorously unitary performance ethos. Mitchell's, then, is a naturalism that literally haunts: ghosts erupt from the seams even as the actors play their characters 'straight'. If naturalist performance is to sign the 'truth' of a character's inner life, naturalism's bodies must fall in line, mime their truth within the lines: fractures disappear so that two (or more) affective lives may become one. This suturing of bodies and lives is, of course, the heart and soul of a well-ordered early modern English household, as well as the essence of contemporary psychological-realist performance in the United Kingdom and North America. But it is also the key to the talking cure, pioneered by Freud and Breuer, and mapped onto stage naturalism at its inception at the turn of the twentieth century.[13]

In *Studies in Hysteria*, Freud shows his readers how completely the talking cure relies on a rigidly conformist body. He *knows* when Elisabeth Von R. has not told him the whole truth, because her body rebels, and her symptoms return:

> As a rule the patient was free from pain when we started work. If, then, by a question or by pressure upon her head I called up a memory, a sensation of pain would make its first appearance, and this was usually so sharp that the patient would give a start and put her hand to the painful spot. The pain that was thus aroused would persist so long as she was under the influence of the memory; it would reach its climax when she was in the act of telling me the essential and decisive part of what she had to communicate, and with the last word of this it would disappear. I came in time to use such pains as a compass to guide me; if she stopped talking but admitted that she still had a pain, I knew that she had not told me

everything, and insisted on her continuing her story till the pain had been talked away.

(qtd in Diamond, 1997b, p. 16)

Freud is the master of Elisabeth's physical suffering; he is in control of her body's capacity to produce evidence. Freud's description is for me a stunning refraction of early modern conduct logic: he guarantees for Elisabeth and for us that her pain may be wholly erased by, absorbed into, the curative promise of language (the promise of the Word of Freud, as science's new God). While this parallel is instructive, it also has obvious limits; I want to dwell here just long enough to emphasize how crucial, and yet finally how forgettable, Elisabeth's body is in Freud's passage. Her pain makes his analysis possible, mimes his expectations – and then it simply goes away. Peggy Phelan (1997, pp. 44–72) offers a remarkable reading of the misplacement of the body in *Studies in Hysteria*, and reminds me that before there was a talking cure there was a physical cure, a clinical commitment to bodily affect within the psychoanalytic apparatus that precedes Freud's obsession with the curative potential of the 'right' words. For Phelan, the talking cure that disavows its debt to bodies will find it cannot compass the flesh it pretends to master: once her treatment has ended Josef Breuer's patient, Anna O, apparently defies her analyst by generating an hysterical pregnancy that implicates him, and specifically his sexualized body, in their work together, while also imagining a corporeal afterlife for that work, residual traces of speech in flesh (ibid., pp. 64–7).

This is exactly the defiance Katie Mitchell's radical naturalism accomplished in her *Woman Killed*. Mitchell materialized the play's controlling discourses, rendering language physical, at turns gorgeous and brutal. She pulled words across her performers' bodies, marking them with language's most violent traces and producing the body of discourse (discourse's expected body) as symptom rather than cure. If Lavinia's hysterical body is an unruly mimic, unable seamlessly to rehearse early modern culture's prescribed truths about a woman's experience of rape, Anne's body in this production arose from hysteria's parallel power to produce excessive meaning. It was hypnoid, literally doubled; its pain escaped the curative of analysis. As Reeves and Maloney demonstrated at Anne's deathbed, Mitchell framed those key moments in the play when the meaning of 'kindness' is up for grabs by projecting the compliant and dutiful body produced by the rhetoric of Christian benevolence alongside the inescapable voice of the suffering body on which that benevolence is staged. Flesh and Word went to war in this production, leaving Mitchell's audiences responsible for bearing witness to the carnage, and to a female subject whose pain cried out despite the tight controls of duty's rhetoric.

From the moment audiences entered the house, through a lighting pre-set that illuminated clearly only the huge wooden cross at centre stage, Mitchell telescoped her battle between flesh and Word through the overwhelming absence of God's own body. The cross, spare and simple, stood as both altar and sentry downstage centre, dominating the rustic thrust space. In Mitchell's imagining, Anne and Frankford's world was arranged around religious rites – around the often awkward, at times openly hypocritical negotiation between Christian doctrine and its daily practice that so vexes both this play and the broader cultural debates on which it depends. The performance opened, closed, and was peppered throughout with ritual singing in an Eastern European tradition, while the characters regularly prayed, singly and in groups, to the cross 'in a manner that recalled medieval or continental Catholicism' (Barker, 2007, p. 167). These rites represented the characters' plainly evangelical uptake of Christian orthodoxy as at once all-powerful, inescapable, *and* spectacularly performative: faith *was* spectacle in this world, and the practice of ritual stood in for and measured the characters' struggles to practise a daily Christian ethic. The actors raised their voices in harmonious unison; the sound, at times very loud, filled the air, overwhelming the playing space. (I found myself surprised and moved by the effect of the sound even on tape.) In these moments of song, text and flesh became one: human voices rendered God present. And yet the singing was not (or not just) designed for edification: it was at once viscerally overwhelming and intellectually disruptive, a vocal *gestus* that encoded a double meaning. The characters often sang to their audience, or to the cross, in formation; framed by this formal presentation style and again by their place within a play, the songs resonated as both worship *and* performance. The sacred music also broke into the script at telling moments, as Mitchell exploited the performative contrast between potentially violent speech and the lyrical sounds of religious devotion. In one memorable instant, Frankford's angry contemplation of Nick's first report of Anne and Wendoll's dalliance was cut short by the entrance of other company members singing at full volume. Frankford's voice was measured as he contemplated his course of action and struggled against his urge to violence; Maloney's trademark sibilant emphasis on 'seem' in Frankford's final line, 'Till I know all, I'll nothing seem to know' (8.115), filled the stage with foreboding and clashed with the sacred harmonies of the cast. Their momentarily too-loud voices literally swept Frankford into the next scene, forcefully erasing the sound of his much quieter, disquieting speech.

Mitchell's massive wooden cross was the focal point of this devout stage world. The physical correlative of voices raised (too) loudly in song, the cross encircled Mitchell's stage with faith's own queer status, God's missing body made tangible on earth not through miracle but through theatre. Even as it stood for the battered and bruised flesh of Christ it also stood for (and by the grace of) the characters' devotion to spectacle, for the tricks of theatrical transformation that let signs be taken for wonders and (any one version of) God emerge from the stage machinery. In its rustic simplicity, Mitchell's cross easily recalled labour – of the characters in this harsh rural world as well as of those in the scene shop who build and rebuild the stuff out of which theatre's magical transformations are made. Aleksandra Wolska reminds me that performance is a plastic art that comes into being 'in the often absurd struggle with the resistance of matter' (2005, p. 93). The material objects that become charmed signs when the house lights dim will be recycled for another production's referents soon enough. The cross is always both prop and icon, the ludic manifestation of the Word of God – as performance.

When Maloney's Frankford passed his judgment against Anne in scene 13, it was to this multivalent icon that he offered his prayers for patience – and addressed Frankford's curse. But the cross's authority could not so easily be borrowed. Because it had already been marked as a harbinger of the performative power of religious language on the material stage – that is to say, of orthodoxy (in this case a worrisomely evangelical orthodoxy) as an always *theatrical* negotiation between the language of doctrine and the practical needs and expectations of its earthly interpreters – in scene 13 the cross did not permit Frankford to escape the scrutiny of his own words (spoken in God's name) and their deeply un-Christian material effects. In this pivotal scene the body in violence, embedded within yet also abject to Frankford's speech, erupted with full and disturbing force. Mitchell and her actors laced Anne's and Frankford's words with physical actions designed to reveal the body *beneath* the paradox that motors Frankford's judgment; these actions literally incarnated the violence *of* the script as at once a form of ritual worship, a form of companionate care, and the exercise of stunning cruelty. 'When do you spurn me like a dog?' Anne begs of her husband (13.93) as Reeves rose to her knees and crawled toward Maloney on all fours, stopping at last to kneel at his feet. Becoming the dog whose treatment Anne craves, Reeves married signifier to signified in a dissonant gesture of supplication that was also a shocking moment of self-effacement: she was wife, she was worshipper, she was animal. She beat herself, as though 'trying to enact the punishments [Anne] imagined' (Barker, 2007, p. 173). Finally, Reeves prostrated herself before Maloney, face down, arms splayed, in a position of extraordinary vulnerability; she seemed as a body already knocked unconscious, but also as a body in prayer, a body made in the image of a cross. Dressed all in white, Reeves's body startled against the larger dimness of the production, becoming momentarily a flesh-and-blood shadow of the cross at centre stage. Tense and terrorized yet graceful and dance-like, Reeves's body posed the question of just whom that icon served, whom it witnessed, and whose bodies it buried under its outsized reflection. Roberta Barker reads these images as evidence of the extent to which Reeves's Anne had internalized the 'moral laws that condemned her' (ibid., p. 173); in the context of this reading it seems to me significant that Reeves should equate her gests of penance (beatings, bestial supplication) explicitly with gests of grace (her prostration before Frankford and the cross, *as cross*). As she would do in the final scene, Reeves's Anne here articulated the cultural script just a little too well: she strove to tame her body into worshipful obedience, but found herself unable to disconnect her call for penance and salvation from the shattering image, and experience, of her body in pain.

Anne's too-literal body became a puzzle on the stage: she at once complied with and defied discourse, and thus it could not contain her. If 'kind' language will not work to tame an errant wife, Whately reminds us, you may choose the 'kindness' of blows; Maloney's Frankford instead chose the threat of rape. He grasped Reeves 'as if to throttle her, then turned her over and threw himself on top of her, yanking one of her legs up' (Barker, 2007, p. 182). After a struggle, both Maloney and Reeves ended up curled in the foetal position on the upstage floor in a tableau oddly, ironically reminiscent of a couple safe and comfortable in bed. This entire physical sequence appeared in counterpoint to Frankford's remembrance of the proper standard of care he had shown his wife: 'Was it for want / Thou playedst the strumpet?' Maloney asked as he slapped Reeves's Anne around (13.108–9); 'Did I not lodge thee in my bosom? Wear thee here in my heart?' (ll. 114–15). When Anne replied 'You did' (l. 115) Maloney's Frankford transformed what had been a momentary loving embrace (as he hugged his wife tightly to his body) into yet another act of violence, gripping harder, squeezing, finally lifting Reeves's body off the floor in struggle. She collapsed; he forced her to her feet to face her children, gripping her from behind, later grabbing her by the neck and throwing her against the upstage wall. Words that seemed to remember love turned visibly, shockingly violent, as every apparent gesture of care choreographed into this macabre dance seamlessly transformed into a parallel gesture of devastating physical harm. When Maloney's Frankford returned from his study to tell Anne that he would neither martyr nor mark her but kill her with kindness, it wasn't the words that surprised; it was the voice that delivered them. As I have noted, Maloney cultivated a low, chilling vocal tic throughout the second half of the performance, and it culminated here as he let his voice become a witness to the unspoken in Frankford's sentence. His pitch rose and fell like singsong, recalling but also making strange the cast's liturgical harmonies elsewhere in the performance. As he reached 'kindness', he placed sibilant stress on the 's', enunciating the entire word with excruciating care and letting the final consonant run. Conventional readings of this play call Wendoll the serpentine tempter, but Maloney's voice seemed to position Frankford in this role. And yet Maloney and Mitchell took care not to set Frankford up as either a plain-clothes devil or a rebel angel. As he moved to the cross for strength and conviction in this pivotal moment, Maloney put Frankford on a collision course with that object's manifold symbolism – as the sign of God in his absence; as the sign of the characters' rigorous, evangelical Christianity; as the sign of the play in performance, religious ritual as play – and located his determination to marry kindness and cruelty within, not against, its powerful orthodoxies of God, doctrine, and theatre.

For Mitchell voice is matter, the physical manifestation of language's own corporeality and the sign of its awesome power over bodies to effect good or to do material harm. Her technique for querying the relationship between language and body through voice is selective amplification. In this production, the loud harmonies of the cast before the cross, Frankford's low, measured, sibilant promise (his curse literalized in breath, with tremendous force), Anne's strained deathbed gasps, Frankford's convulsive crying at her side, and the flat, emotionless testimony of Anne's friends and relations all competed to witness the dissonant relationship the performance enacted between ostensibly Christian 'kindness', household care, household violence, and the promise of salvation. The voice Mitchell invoked for *Woman Killed* is not, however, the voice of earlier feminists who have raised the radical potential of aurality against the tight strictures governing visual representations of the female body; rather, it cathects with Judith Butler's recent work on Levinas's face. Butler reads the 'face' as word and image confounded: it is not a human face at all but a face turned away, a back turned toward us. It is 'a series of displacements' that finally allow

Levinas to imagine 'a scene of agonized vocalization', 'a figure for what cannot be named, an utterance that is not, strictly speaking, linguistic' (Butler, 2004, p. 133). The face undermines the salvation narratives that haunt the women in this chapter. It signs the Christian prohibition against murder – 'thou shalt not kill' – and yet it may not be assimilated seamlessly into doctrinal interpretations of God's Word, for it speaks in excess of Symbolic language, speaks 'an agony, an injurability, at the same time that it bespeaks a divine prohibition against killing' (ibid., p. 135). 'The face, if we are to put words to its meaning, will be that for which no words really work', Butler writes. '[T]he sound of language evacuating its sense', the face is the sound of human suffering staged in front of (in the face of?) the call to translate that suffering into some other kind of sign (ibid., p. 134). It is the voice not before language but beneath it, expelled by it, the agonized sounds language disavows in order to constitute itself as the mark of civilization. The face is thus also, and most importantly, a call to ethics (p. 134), to an act of witness that reads between the signs, that reads the limit of the sign's power to represent an other.

In her 2004 staging of Euripides's *Iphigenia at Aulis* for the Royal National Theatre, Mitchell used the same technique of selective vocal amplification to very similar effect, and the comparison with her earlier *Woman Killed* proves for me instructive.[14] Mitchell set *Iphigenia* on the Mediterranean holiday coast *circa* 1940. She foreshortened the long Lyttelton Theatre proscenium; a narrow but airy room ran length-wise across the stage and most exits and entrances were made stage left, through an area invisible to the audience but designed to give us the impression of a plaza, the sea, and a mass of Greek residents assembled to see Menelaus and Agamemnon sail for Troy. Agamemnon lures his wife Clytemnestra and his daughter Iphigenia to Aulis under the pretence of the latter's wedding to the eligible Achilles; the true reason for their journey is Iphigenia's sacrifice to bless the war at Troy. The performance climaxed after Clytemnestra (Kate Duchêne) and Iphigenia (Hattie Morahan) desperately but unsuccessfully pressed Agamemnon (Ben Daniels) for a reprieve. Realizing she would have to face death as a gift to the gods, Morahan's Iphigenia began to sob uncontrollably, first reaching toward Daniels and later collapsing to her knees with Duchêne. Iphigenia and Clytemnestra sang 'a last song' together through their convulsive, panicked tears; the song was one of cheerful journey, punctuated by the loud gasps and heaves produced by Morahan and Duchêne's overwhelmingly physical grief. Their bodies' voices – gasps and heaves, hysterical panic – warred with the text of the song. Iphigenia was no mythic sacrifice, or not yet; at that moment, she was a frail and frightened girl.

Then, all of a sudden, the tone changed: cold white lights flooded the stage, and Iphigenia made an about-face. 'I must die, and do it with dignity', Morahan told us, shouting the line at Clytemnestra above the humming din of the advancing army's machinery. Like Anne on her deathbed, Iphigenia attempts to reclaim control over her experience and articulate the promise of salvation (framed as the heroic glory of a pre-Christian nation, the salvation of the state rather than the individual) in the face of devastating and unjustified violence against her person; here, Morahan's tool for this reclamation was a microphone the chorus women brought on stage. Again, as in *Woman Killed*, Mitchell's actors produced 'too much' naturalism for comfort: Morahan fumbled, panicked, as she tried to turn on the microphone; every actor on stage was wired, jumpy, ready to explode as their affective and intellectual responses to the situation struggled within their bodies. 'I dedicate my body as a gift for Greece', Morahan finally said, measuring each word into the microphone; 'Take me, sacrifice me, and then to Troy. Plunder the whole city, and when you leave it, leave a ruin. That will be my memorial.' She stepped back, momentarily unable to continue; she took her hands off the microphone stand, which she had been clutching for support, and pressed them hard into her eyes. Spectators watched her struggle openly to find both language and the tongue to speak it; at several moments she broke down, gasping and crying, body falling forward to keep her head below the microphone's mouthpiece. We heard her heroic speech for Greece, but we *also* heard the strangely familiar, utterly excruciating sounds of a very human body trapped in horrific circumstances.

In Morahan's panic, her strain to make the speech of a lifetime, I heard the echo of the missing bodies beneath the pristine deathbed testimony of women like Anne Frankford and Katherine Stubbes. And, even now, I find the gasping, gulping voices of Reeves's Anne and Morahan's Iphigenia stay with me; I am haunted by the memory of their failing, flailing attempts to will their suffering away and wrench their bodies into line with the performances of sacrifice and salvation demanded of them. Both Mitchell's *Iphigenia at Aulis* and her *Woman Killed With Kindness* ask us to witness the process by which violence against a woman dissolves into an exemplary 'kindness' (for husband, for household, for nation), but they also ask us to understand what that process does to the bodies expected to carry out the extraordinary task of erasure. For Reeves's Anne as for Morahan's Iphigenia, the show of the cover-up was its own incredible blow: the taming of the suffering body into the body of language, of command performance, was harder for me to see, to hear, and to witness than were the (far more conventional, far more expected) acts of other hands that preceded and followed. Mitchell's work in both productions forced me into affective, visceral collisions with characters, situations, and ideas for which I was not prepared, but which, in their perpetual haunting, help me to imagine what it might feel like to encounter the violence of violence's cover-up, the receding sounds of suffering within another's body. In the next chapter that encounter will become my main focus as I tell the story of one sacrificial object who openly refuses to play by salvation's rules.

4
Witness to Despair: The Martyr of Malfi's Ghost

> Ghosts hover where secrets are held in time:
> the secrets of what has been unspoken, unacknowledged;
> the secrets of the past, the secrets of the dead.
> (Rayner, 2006, p. x)

Witnesses and ghosts

The performance of salvation produces a domestic martyrdom: it obscures the bodily and psychic destruction wrought by household violence, positing union with God as a battered wife's happy reward. Household becomes scaffold as the quietly suffering wife becomes her husband's spiritual exemplar; her quiet submission to violence becomes proof of her commitment to the integrity of his household and family, devotion to which stood as a cipher for devotion to God in post-Reformation England (Peters, 2003, pp. 291–2; Owens, 2005, pp. 107–8). Such stoic – even apparently pleasurable – suffering, moreover, gave a wife like Anne Frankford the temporary authority to speak freely, to testify to her religious commitment in a semi-public way. Nevertheless, as the final moments of Mitchell's *Woman Killed* remind me, the performance of salvation is always a precarious act, forever haunted by the violence it forecloses and by the tension, ambivalence, terror, and doubt that violence trails in its wake.

Anne Frankford dives with relish into the task of staging her domestic martyrdom, but her failure fully to mask the physical and emotional losses that attend it represents an alternative proof of faith. In this chapter, I want to take Anne's story a step further as I examine a deliberately recalcitrant martyr figure, Webster's Duchess of Malfi. In contrast to Anne's paradoxical violation, shaped by the *fort/da* operations of 'reasonable correction', the source (Ferdinand) and nature (beyond conscionable) of the Duchess's torture are never in question in Webster's play. And, unlike the compliant Anne, the Duchess loudly, brashly, angrily refuses to stage-manage her body's destruction, refuses to disavow the rage, pain, and loss with which she struggles. The threat of violence's unwieldy return does not mark the outer edges of the Duchess's performance labour, as it does Anne's and even Lavinia's; rather, that threat is central to it. The Duchess is salvation's scourge, howling relentlessly the obscured links between violence against her person and the promises of comfort, grace, and peace with which Bosola attempts to placate her; in the process, she not only ruins the spectacle of salvation but also openly challenges those who would bear it, and her, witness.

Post-Reformation martyrologies image the female martyr as a 'godly wife', Protestant England's devotional icon; she idealizes the familial bonds of care that shaped the post-Reformation nation-state and defined its subjects. Female Protestant martyrdom was an act of nation building, of social subject-formation that depended upon a religious woman's ability to perform devotion in the face of public suffering, but in such a way that masked both performance *and* suffering. Importantly, the physical destruction that is an integral part of the experience of martyrdom carried very little currency for women: while the martyrdoms of their male counterparts often pivoted on the difficult tension between the call to faith and the brutal loss of life embedded in the moment of execution, for a woman on the scaffold or the pyre the wholesale abdication of her earth-bound body was the only viable proof of faith. This is the double standard the Duchess of Malfi will not accept. She will not be the Protestant model of quiet wifely obeisance, but nor will she abandon her right, as a Protestant subject, a Christian subject, to productive doubt – to a body lived within complex social networks and pervasive religious and cultural uncertainties. Instead, she will be, like Heywood's Nick, the witness who questions, who pulls at the threads linking sacrifice, subjectivity, statehood, and sexed and gendered embodiment.

As the Duchess destabilizes early modern England's all-important links between femininity, faith, and broader cultural production she also raises key problems of reception, both for her original audiences and for contemporary spectators. If the Duchess will not witness her own body's dissolution into faith, what acts of witness does she provoke in turn? Are audiences willing to receive her acts of resistance *as* resistive, to make themselves vulnerable to her refusal to model proper religious subjectivity even as she holds fast to her right to believe in a benevolent God? The Duchess dies, famously, at the end of Act Four: what happens to the legacy of her resistance in Act Five, after she is dragged from the scene? How are her refusals to play along with either Ferdinand (her household tyrant) or Bosola (her jailer and self-styled protector) received, mitigated, and finally recuperated by those responsible for her murder? And how does that recuperation colour the way contemporary actors, directors, audiences, and scholars understand and perpetuate the story of the Duchess's fabled, heroic death? *The Duchess of Malfi* stages the politics of watching women's martyrdom in a cultural moment when watching women in any public display was a fraught endeavour. The play asks its audiences what willing blindness it takes to manufacture a female martyr, and what witnessing her sacrifice could and should mean in the face of such collective cultural aversion. A politicized *Duchess*, in turn, depends upon our willingness to examine the paradoxes that adhere to the Duchess's performance of martyrdom, to interrogate the means by which she is, despite those paradoxes, reclaimed for posterity as unproblematically faithful, and to consider the wider consequences of that reclamation.

When Frankford says that he will not martyr Anne (*Woman Killed*, 13.154), he suggests that such an act would risk granting her further authority in his household and over him. The role of martyr was indeed one of the few in early modern England that could be genuinely empowering for women; as Christine Peters notes, there was something of the rebel in each Protestant female martyr, all the more reason for the narratives that celebrated her sacrifice to reimagine her as a model wife and supplicant after the fact (2003, pp. 291–2). This small detail is important: while female martyrdom may have depended for its social currency on the woman's immediate performance of salvation in the moments of her torture and final, public execution, that martyrdom was ultimately defined and controlled by her performance's posthumous, third-party textual transcription and dissemination. As I suggested in Chapter 3 in relation to both Heywood's Anne and Mitchell's *Iphigenia*, the accidental return of violent sacrifice's rough and painful margins is an almost-inevitable consequence of the instability, immediacy, and improvisational quality of live, embodied spectacle. The threat of that return makes the posthumous rewriting of a martyr's 'last public performances' (Zimmerman, 2005, p. 62) in a more fixed and containable medium absolutely essential in order to control their broader signifying power. Which leaves me wondering: if even a standard performance of salvation such as Anne Frankford's risks pulling martyrdom's script off track, how much more urgent must be the recuperation of one, like the Duchess, who tears that script to pieces, raises her own ghosts and gives them parts to play?

The Duchess of Malfi thus stages a fresh and particularly urgent version of the in/visible act. The play centres on the erasure not of the Duchess's violation – which we all see, pity, and mourn – but of her confrontational, interrogatory performance of rage and retaliation in response to that violation. Similarly, the play foregrounds not the cruelties embedded in the performance of salvation, but the risks, dangers, and critical possibilities inherent in the acts of spectatorship that performance invites. Lavinia and Anne Frankford both experience the effacement of their suffering as a product of the containing and normalizing performances to which they are enjoined, only to disrupt those performances with the unexpected eruption of bodies that cannot so easily be made whole or sensible. In both cases, it is the trauma of violation – rape, household 'correction' – that performance elides, and the trauma of elision that the in/visible act exposes as the violated body wedges itself between the script and its articulation. In *The Duchess of Malfi*, by contrast, the Duchess's deliberate acts of resistance are the *object*, rather than just the menace, of erasure – violence doubly marginalized. In Webster's climactic Act Four, the Duchess bears provocative witness to the performative and discursive practices that would obscure her suffering body and the complexities of its experience in social and affective space; as her death scene gives way to an Act Five in which she reappears only as a ghostly echo, those acts of rebel witness are rewritten as the emblematic death of a model godly wife. Act Five styles the Duchess as heavenly protector of both family and nation, rendering her earlier resistance futile and her legacy 'reactionary': 'she is removed from the potentially radical conflicts [...] that have fully defined her, and the sympathy and value assigned to her life are unambiguously allied instead with a compelling tribute to the lost past' (Rose, 1988, p. 172). The Duchess, martyrdom's conscientious objector, becomes fully domesticated, literally transformed into the ghostly remnant of Duchess-as-martyr.

In this chapter, I try to imagine what might happen if feminist critics, practitioners, and audience members were to regard Webster's fifth act, and particularly the famed 'echo scene' (5.3), not as an emaciated denouement to the Duchess's far more compelling life and death, but rather as an opportunity to stage precisely what goes missing when Bosola attempts to resurrect his mistress-cum-prisoner as a supplicant to God's 'mercy' (4.2.352) – and as a model for his own spiritual reinvention – in the eulogy he delivers over her dead body.[1] As I argued in my introductory chapter, the in/visible act is foremost a call to witness one's own acts of watching: it throws the processes of percepticide into relief and invites an interrogation of what Taylor calls ' "just looking," dangerous seeing' (1997, p. xii). In these terms *The Duchess of Malfi* telescopes the in/visible act's very conditions of possibility, offering essential commentary on how we encounter – and *fail* to encounter – women's own responses to the marginalization of their violated bodies. If Act Five leaves the Duchess almost completely out of the frame – permitting her, in her cameo return as Antonio's echo, small voice and no body – then Act Five must become nothing less than a call to witness 'beyond recognition' (Oliver, 2001), to remember the absence of the Duchess and read that absence as a spur to reconsider what and how we see (of) her, and what and how we 'saw before' (Blau, qtd in Rayner, 2006, p. xviii).

An echo is a sound negative: speech in the act of disappearing, the sound of sound going missing, the sound of the origin as unstable ground. For early modern vocal historian Gina Bloom, 'echoic sound' 'disrupts the unity of voice, body, and subjectivity' necessary to normative (what Diamond would call patriarchal) mimesis (Bloom, 2007, p. 163). The echo is the aural correlative of the in/visible act, the shadow of a voice already spectral to itself, and it carries with it the power to infect listening ears, to reverberate watching bodies, to implicate those bodies physically as well as emotionally in the story of its disappearance – a story itself on the cusp of fading away. We encountered something of the power of the echo in the last chapter, in Saskia Reeves's rasping voice and Hattie Morahan's convulsive sobs. But the Duchess of Malfi's Act Five echo is – could be – an even more radical sound negative. It haunts not the performative acts that would efface violence against women, but the acts – performative, textual, spectatorial – that would obscure any trace of women's attempts to examine, expose, or resist the processes of that effacement. This ghost of the Duchess's ghost haunts the given circumstances that frame spectacle's relationship with its spectators, pointing always beyond the immediate scene of representation and toward 'that which cannot be reported by the eyewitness, the unseen in vision and the unspoken in speech' (Oliver, 2001, p. 2). It thus bears the capacity to transform a play infamous for its morbid, fetishistic specularity into a play about acts of critical witness – into a site for modelling a spectator who pushes beyond the specular and into the space of ethical encounter beyond.[2]

The radical echo I envision for a feminist performance of *The Duchess of Malfi* need not be constrained by act and scene borders. In fact, the two productions I explore in this chapter – Phyllida Lloyd's for the Royal National Theatre (2003) and Peter Hinton's for the Stratford Shakespeare Festival (2006) – each in their different ways suggest that *The Duchess of Malfi* as a whole is haunted by what it elides of the Duchess's response to her pain and its marginalization. These productions throw the play's collisions between spectacle and spectre into relief, staging a sustained encounter among character, performer, witness, and the ghosts of the missing lying between them. Echoes proliferate in Hinton's production: bodies ghost bodies, the Duchess never appearing in the singular, never appearing but as a palimpsest of fragments, a figure both haunted and haunting. Lloyd, meanwhile, scatters her playing surface with mirrors up to audience, challenging her spectators to imagine our own bodies within the Duchess's visual field, to experience watching the Duchess as a subset of watching ourselves, to experience watching ('just looking') as an act heady with personal as well as political implications. Hinton's Duchess will not be confined to one performance body, will not (cannot) exist without a witness; in Lloyd's production, we become the shadows, the echoes, the Duchess's ghosts. Together, these two productions return the Duchess's dissenting figure not only to the stage but to the stalls, stage the ghost of Malfi's ghost as the model witness to Malfi's own, elided performance of critical despair.

The Duchess dies a martyr

> Let Heaven, a little while, cease crowning martyrs
> To punish them.
> Go, howl them this: and say I long to bleed.
> (4.1.107–9)

The substantial body of feminist scholarship on this play has long been preoccupied with the question of what kind of martyr the Duchess of Malfi turns out to be. More traditional work reads the events of Act Four as the inevitable transformation of a rebel ruler into a 'stoical Christian heroine' (McLuskie, 2000, p. 112). Christy Desmet, echoing Lisa Jardine's early and influential comments on the topic (1989, pp. 71–2), calls the Duchess a 'strong woman' who quickly becomes 'a comfortably passive figure of female suffering', finally dying 'with a martyr's [...] acquiescence' (Desmet, 2000, pp. 50, 53). Theodora Jankowski argues that Webster ultimately undermines the prominence of the Duchess's political body with a 'final representation' of her 'not as ruler, but as idealized suffering wife/mother/woman' (1992, pp. 178–9; see also L. Hopkins, 2003b, p. 28). Other critics have keyed the Duchess's martyrdom to her status as what Frank Whigham calls a 'family pioneer' (1996, p. 201): in this scholarly thread, the Duchess is martyred either to her passionate attachment to her family – a distinctly Protestant act of worship (see Peters, 2003) that directly opposes Ferdinand's insistence on ancestral hierarchy and blood purity (Rose, 1988, p. 171) – or to her sexual independence, her proto-feminist demand for parity of desire (Daileader, 1998, p. 80). In both paradigms, martyrdom is the play's response to the Duchess's sexual and social transgressions, and she is, however perhaps unwillingly, read as finally complicit with the act of containment it represents.

In citing these readings of the Duchess as martyr, I do not wish to imply that they are without merit. The Duchess of Malfi is a challenging figure in part because she appears – even as she rails against the God(s) of Bosola and her brothers; even on the cusp of her murder – to be a devoted Christian. She also fulfils a number of key criteria for Protestant martyrdom in Act Four despite her simultaneous hostility to the idea. She faces the machinations of an outrageous tyrant in the form of Ferdinand and his (very Catholic) ritual displays of dead bodies and relics; though she rages against his 'horrible' (4.1.53) spectacles she also resigns herself stoically to 'endure' what she openly calls his 'tyranny' (4.2.60). Her clamouring at the trials of 4.1 gives way to the quiet contemplation of 4.2 (ll. 29–30), and, as Ellen Caldwell argues, this structure (a boisterous examination scene, followed by a stoic death scene) needs to 'be placed within the context of early modern institutions of examination', including 'religious inquisition' (2003, p. 150). The Duchess's calm, rational reception of both Ferdinand's madmen and Bosola's turn as masked tomb-maker in 4.2 suggests a victim of persecution who has mastered the Reformation doctrine of 'passive obedience', the same religious/political rule that governs the behaviour of the battered wife and 'enjoin[s] subjects to adopt an attitude of patient endurance until such time as the mysterious workings of divine justice [bring] about the downfall of the corrupt ruler' (Owens, 2005, p. 89; see also Bevington, 1968).

And yet for every gesture the Duchess makes toward a martyr's calm, confident faith – 'You violate a sacrament o'th' church / Shall make you howl in hell for't', she tells Ferdinand (4.1.39–40); she insists her coffin 'affrights not me' (4.2.171); she forgives her executioners (4.2.206) – she makes another that mocks the pose of the heroic female supplicant, the quietly noble godly woman into which Bosola's doctrine would transform her. 'I have so much *obedience* in my blood, / I wish it in their veins, to do them good' (4.2.168–9, my emphasis) the Duchess tells Bosola when he brings coffin, cord, and bell. Her sneering 'obedience' is no less a curse than Frankford's queer promise of 'kindness' is to Anne: it is an infusion of poison ('I wish it in their veins') that briefly exposes the underside of early modern orthodoxy, its yoking to household and state cruelty. Throughout Act Four the Duchess's martyr's pose abuts her 'strange disdain' (4.1.12); she is never simply hostile, nor is she simply supplicant, but she openly mixes the two with a peculiar and wicked irony. Finally, the Duchess is left uncertain about her relationship to the Church, if not to God himself; her faith as a Christian is unmoored by her acute awareness of faith's gendered double standards on earth (Whigham, 1996, pp. 210–11). As my (slightly manipulated) epigraph to this section suggests,[3] she is fully alive to the subtext of Ferdinand's punishment and its broader consequences. Heaven should cease crowning martyrs, she rails; heaven should instead punish them, should acknowledge that martyrdom is – in this case, in all cases – *also* brutal punishment, that Bosola's quiet entreaties of spiritual comfort ('Come, be of comfort; I will save your life' [l. 86]) are the soft underbelly of Ferdinand's blasphemous tyranny (l. 116). Heaven should spell out its intentions, cries the Duchess, rather than 'wrap' 'poisoned pills / In gold and sugar' (4.1.19–20). For, just as the poison-wrapped pill will taste sweetly but wrack the body, the Duchess argues that she can be no heavenly martyr – can be no Christian subject – without *also* being a woman on earth in rage, doubt, and pain.

In a nuanced essay on the material artefacts of early modern household life, Wendy Wall reinvigorates existing debates about the Duchess's relationship to privacy, domesticity, and corporeality by exploring the several ways in which foodstuffs and home remedies circulate in the play. She reads the Duchess's death scene thus:

> Mocking Bosola's suggestion that strangulation is an exceptionally grim way to die, she asks if a perfumed smothering with cassia (another element found in home remedies) would be preferable; that is, she refuses the false sweetness of death and insists on its quotidian but transformative production of remainders. [...] [She] raises the possibility of dying in the state of sugar-sops oblivion [...] only to renounce it: that is, in having violence substitute for ('serve for') mandragora, she pointedly refuses a drugged ending and declines to die domestically in the conventional sense of the term.
>
> (Wall, 2006, pp. 170–1)

Wall implies that the Duchess's death is notable not for its apparent 'false sweetness', in which she forgives her executioners and turns at the last to the continued care of her children, but because even in her final moments she acknowledges, and makes a dark joke from, the material, social, and ideological 'remainders' that shape the world in which she has lived: a world in which cassia may be a sweet, a curative, and a source of violence all at once; a world in which she may be a devout Christian and yet violently resist early modern Christianity's cruellest tenets; a world where she may be martyred and yet may absolutely resist martyrdom even in the act of it.

My reassessment of the play's representation of this martyrdom takes its cue from Wall's careful historicization of the Duchess's domesticity, her recognition that the Duchess can be a figure of *productive* ambivalence for contemporary scholars and practitioners. In order to understand the play's construction of the Duchess in Act Four and her equally vexed, even vexing response to that construction, I read the Duchess in relation to the scripts of both male and female martyrdom in the sixteenth and seventeenth centuries in England, as well as in relation to the scripts of household suffering and salvation that I investigated in Chapter 3. I frame the Duchess of Malfi as a product of the several social narratives that shaped the 'godly wife' as religious exemplar, a body abidingly passive and yet nevertheless in possession of world-making public power. And I consider how all of these narratives finally try to resolve their competing articulations, figuring the godly wife/female martyr as a text to be read in posterity *without* contradiction.

Bodies, acts, texts

At the beginning of Chapter 3 I told the story of Katherine Stubbes, a devout Protestant who apparently willed her own death, staged the model post-Reformation performance of salvation and created an ideal precedent for Anne Frankford's eerily similar end. In 1591, Phillip Stubbes published the story of Katherine's death as *A Crystal Glass for Christian Women* and claimed on its title page to have 'set downe word for word' Katherine's deathbed sermon 'as neere as could be gathered' (qtd in Phillippy, 2002, p. 89). This claim will likely strike modern ears as suspicious, but it was clearly attractive to Stubbes's contemporary readership: *A Crystal Glass* had gone through 24 editions by 1637, and as many as 34 by 1700 (Klein, 1992, p. 140; Kidnie, 2002, p. 5; Phillippy, 2002, p. 81). Both a formal exercise in funerary ritual as well as a classic Protestant martyrology, *A Crystal Glass* is, as I noted in Chapter 3, ultimately far more reflective of Stubbes's religious and social preoccupations than of the 'real' life and death of his wife, especially in the lengthy and sophisticated deathbed testimony it attributes to her (Phillippy, 2002, p. 94). The Katherine I invoke above and animate in Chapter 3 is thus the Katherine not of private history, but of public legend; her story is the work of Stubbes himself, while the diva presence he conjures neatly effaces his role as editor, director, creator.

In her extensive reading of Katherine's deathbed scene, Patricia Phillippy argues that, for all their ritualized convention, tracts like *A Crystal Glass* worked as public acts of private mourning (2002, p. 92). While this text may indeed represent something of Phillip Stubbes's own grieving process (McLuskie, 1989, p. 37), it is also a clear manifestation of his power and authority as head of Katherine's household, as the man ultimately in charge of shaping that household's public image. Katherine's final days may or may not have been as seamlessly 'godly' as Stubbes's account suggests; we have no way to know, because Stubbes's rendering of Katherine's suffering and salvation admits no uncertainties. If she was hurting, if she was angry, if she lost faith at any time, he does not reveal it. Stubbes's narrative becomes the *sine qua non* of Katherine's deathbed performance: his scripting, after the fact, of that performance *as* idealized, *as* seamless, *as* godly renders it for posterity without any word or gesture out of place. Katherine's ritual enactment of her faith in some form before her death would likely have been enabling for the text of her husband's subsequent martyrology, but its authority is wholly dependent on his intervention. Her ephemeral performance in her final days is essential, yet entirely subservient to the script that follows: it is always already a product of that script.

Feminist scholars have often noted that women's bodies occupied an awkward position in the early modern cultural imaginary: they were central to, and yet also had to be *forgotten* as central to, a host of public acts, from funerals (Phillippy, 2002, pp. 15–48, esp. 28–9) and regular religious observances (Peters, 2003, pp. 154, 178, 196) to mothers' advice books (Wall, 1993, pp. 285–7) and, as we saw in Chapter 2, even the reporting of rape. This systematic erasure of the ritual performing body makes perfect sense in light of the culture's anxieties about women's symbolic authority; as Frances Dolan argues, because women's bodies 'played important roles in defining and securing masculine power', to show those bodies vulnerable in public space 'would [have] undermine[d] masculine authority and privilege' (1994b, pp. 166–7). But it also makes sense in light of what Alice Rayner identifies as the awkward relationship among text, performance, and traumatic representation: '[w]ritten history is generally a writing away from trauma', she argues. 'Performance, on the other hand [...] speaks and embodies the specific relatedness out of which the trauma arose in the first place; it therefore works toward trauma' (2006, pp. 26–7). Performance, in other words, is the medium of rupture: it relies on the body, and it calls deliberately upon the bodily repressed.

Early modern martyrdoms, both male and female, turned on the repression of bodily care in face of the performance of salvation through faith on which Protestant doctrine rests. Catharine Randall Coats describes the privileging of text over body as one of the central tenets of Protestantism (1992, p. 12), while Cynthia Marshall makes a similar point in relation to Foxe's *Actes and Monuments*, the period's most widely read martyrology. 'In Foxe's theology of the word, textuality subverts the force of physical presence'; reading becomes 'a godly act' as the stories of destroyed martyr bodies '[participate] in the development of the individual as reader and the reader as individual' (2002, pp. 96, 88; see also Zimmerman, 2005, pp. 63–4). Men's bodies, like women's, were made to perform their own erasure in the moment of execution (Zimmerman, p. 65), but the specific anxieties framing the public appearance of women's bodies in the period, coupled with the formal position women occupied as (domestic) exemplars of obedient worship in English post-Reformation tradition, rendered their experiences of martyrdom especially unstable. In her comprehensive examination of women's religious roles after the Reformation, Christine Peters calls Protestant martyrdom a process of 'becoming male': the Reformation shifted its focus 'from the experience of martyrdom itself' that had characterized Catholic immolations to the 'experience of imprisonment and interrogation' (2003, p. 273) in which the emphasis lay on the martyr's ability to spar with tyrant interlocutors. While the male martyr had cultural leave to interpret scripture and thus to dispute at will without negative social repercussion, the female martyr's simultaneous domestic position as a godly wife/godly woman troubled her powers of public speech and persuasion. She drew her religious authority not from a public talent for scriptural hermeneutics but from her household and parish role as a 'naturally receptive vessel' for scripture (ibid., p. 196) – one who kneeled to listen but rarely spoke. To argue openly with her captors would be to betray her position as passive community exemplar and obedient household mistress, a position on which her larger public persona absolutely depended.[4] The very foundational quality of this dilemma meant that the Protestant female martyr was always awkwardly *in* her body, shaped and constrained by the paradoxes that framed her dual status as vessel and agent, humble supplicant and rebel warrior, until the very end of her ordeal. A figure of domestic supplication whose status as such could only with difficulty admit the vocal and specular contours of exemplary public religiosity, the female martyr risked gender upheaval simply by doing what was expected of her: publicly staging her faith.

On the scaffold or pyre, the male martyr's dying body could be received and recorded in exacting detail (Knott, 1993, p. 79; Dolan, 1994b, p. 161) because the story of his physical suffering acted as a cultural flash-point for the ambivalence and doubt characterizing both Protestant devotion and Reformation subjectivity. The female martyr's body in violence, however, risked making a spectacle of its governing contradictions, making her performance of salvation less a model than a potential site – like the wife's battered body – at which the inconsistencies and violent elisions that buttress patriarchal self-fashioning become uncomfortably visible. Foxe's *Actes and Monuments* offers a telling example of the resulting representational disparities in its separate handling of the burning of Bishop Nicholas Ridley in 1555 and that of Anne Askew in 1546. Both Ridley and Askew were figures of high esteem among the Protestant faithful, and both therefore command lengthy narratives in Foxe's text. Their execution scenes, however, could not be more different. Foxe devotes almost a full page to the minute details of Ridley's burning, which went awry when the wood on the pyre refused to combust properly. As the wood appears to flame Ridley commends his spirit to God, but as soon as the problem with the fire becomes apparent Foxe's account grows ambivalent. Ridley calls on God's mercy, but Foxe also records him clearly struggling, 'intermedlynge' his cries for mercy with the cry 'lette the fyer come unto me, I cannot burne' (1563, p. 1378). Finally a bystander manages to fan the flames and Ridley works himself over to the hot side of the pyre, where at last he dies. Foxe ends his account in the crowd, noting the immense physical affect of Ridley's witnesses, their tears and profound grief (ibid., pp. 1378–9). In Askew's case, Foxe tells his readers that she was brought to Smithfield in a chair and tied to the pyre with a chain, her body having been broken on the rack, but offers no other clues (save that she apparently remained unwavering in her faith) to her physical or emotional state. His remarks on her death are similarly devoid of detail:

> Thus she being troubled so manie maner of waies, & having passed through so manie torments, having now ended the long course of her agonies, being compassed in with flames of fire, as a blessed sacrifice

unto God, she slepte in the Lorde [...] leaving behind her a singular example of Christen constancie for all men to folowe.

(Foxe, 1563, p. 680)

Frances Dolan notes that Foxe's plain account of Askew's death contrasts markedly with Askew's own vivid descriptions of her torture under examination, which Foxe quotes at length earlier in his narrative. Citing Karen Newman and Catherine Belsey, Dolan goes on to argue that the scaffold or pyre was one liminal site at which women's speech was licensed, and was, more importantly, likely to be written down and recorded (1994b, p. 158), but it was also, as in Askew's case, the site at which a female martyr lost any power to control the shape her final experiences would take on the public, political stage (p. 161). Dolan concludes: 'On the scaffold [...] [w]omen are constituted as subjects who think, speak, and act on the condition that they are represented as transcending' their bodies and embodied experiences (ibid., p. 159); the post-salvation narrative that guarantees their 'spiritual triumph' also 'denies their mortality and consequent loss' (p. 165).

In complement to Dolan's work on the discursive erasure of the female body in martyrdom, Deborah Shuger explores the symbolic labour accomplished by the suffering male body in post-Reformation culture. Although English passion narratives are relatively rare, Shuger nevertheless presses their cultural significance (1994, p. 89). She suggests that the 'black and blue' body of the tortured Christ records 'the destruction of manhood [...] naked and powerless' (ibid., pp. 95–6); that tortured body then becomes both an object of identification for the persecuted Protestant elect as well as important physical evidence of social and political injustice against them (p. 97). As the exemplary post-Reformation body, Christ is defined by his shockingly visible suffering as well as by his internal, psycho-spiritual battles. In his iconic struggle between the urge to cause pain and the even more 'urgent demand for self-control, moderation, and obedience to external authority' (ibid., p. 106), the violated Christ enacts the 'self-divided consciousness' (p. 101) of the ideal Protestant subject. He images that subject as an explicitly male body in violence, but also as a body *in doubt* about the meaning of its violence; the model of identification he produces depends upon the recognizability of his own body's physical pain as well as on the complexity of that pain's psychical and spiritual uptake. Subjects do not simply identify with Christ's violations; they identify – as Nicholas Ridley's witnesses clearly did – with the manner of his suffering, and they struggle with its meaning.

Celia Daileader calls the Duchess of Malfi a Christ figure and Webster's text 'a kind of proto-feminist passion play [...] haunted by the dream of resurrection' for the Duchess as a 'female sexual martyr' (1998, p. 82). Daileader's argument is provocative, but it does not fully account for the different conditions of embodiment that govern Christ as exemplary imago and the female martyr as exemplary supplicant, object rather than subject of Protestant patriarchal modelling. The former's body in violence is a site of productive anxiety, but the latter's must abandon its materiality precisely because the anxieties it invokes are *not* patriarchally productive: the parallels between the leaky body of Christ and the leaky feminine body end, necessarily, on the cross. In an extension of Daileader's work, I propose that the Duchess is not (simply) a Christ figure, but that she deliberately troubles the gendered division of labour Christ's allegory sets up when she insists on experiencing her woman's body – its pain, doubt and despair; its ambivalent relationship to heaven; its overwhelming rage and loss – as culturally valuable in the moment of her forced martyrdom.[5] Numerous critics have remarked on the Duchess as an androgynous, even male-identified, figure: she would be ruler over herself and her lands, woos Antonio like a man, controls her own sexual choices, and meddles with the line between public and private identity (Jankowski, 1992, p. 173; Behling, 1996, pp. 27, 33; Whigham, 1996, pp. 202–8; Haber, 1997, pp. 146, 149; Daileader, 1998, pp. 79–87; Rose, 1988, p. 164; Peterson, 2004, p. 259; Barker, 2007, pp. 63–5). But the Duchess is not an exemplary male martyr despite her female body, nor is she the compliant echo that female martyrdom demands. She appropriates and reconfigures for her own experience of violation and loss the doubt, the insufficiency, and the difficult embodiment that characterize Christ's emblematic suffering precisely because those same conditions also haunt, wordlessly, the invisible edges of female martyrdom. The Duchess demands the use of her body as both a pulpit and a stage and, as she rages the story she knows is about to go missing, she inaugurates a new and unexpected Protestant performance economy, one that refuses to bow to the reconstructions of text.

Performing despair

Webster organizes Act Four of *The Duchess of Malfi* around Protestant martyrdom's two central scenes: the martyr's examination (4.1) and her execution (4.2). Likewise, the Act features parallel performance economies. Ferdinand's spectacles – the waxwork dead man's hand and tableau vivant of bodies representing the Duchess's dead husband

and son; the chorus of madmen sent ostensibly to 'cure' the Duchess of her 'deep melancholy' (4.2.40–4) – oppose and are opposed by Bosola's relentless attempts to encourage the Duchess to take his comfort, leave off grieving, put on 'a penitential garment' and take up 'beads and prayerbooks' (4.1.119–21) in a stock performance of supplication. Karin Coddon and Andrea Henderson have separately argued that the play records a transitional moment in English history when older forms of public display pushed up against new ideas about private interiority; especially in Act Four, they claim, theatricality aligns itself with tyranny and domination and against the emerging modern bourgeois subject (Henderson, 2000, p. 62; Coddon, 2000, pp. 39–40; see also Diehl, 1997, pp. 182–212). Henderson and Coddon register well *Malfi*'s marriage of spectacle and punishment, but they do not fully distinguish between spectacle and performance in the play. While the former, as I have already suggested, ranges around ritual practices that register as Catholic (consider Ferdinand's trick with the dead man's hand, which operates as a kind of macabre saint's relic [Diehl, p. 184]), the latter frames the Duchess within the world of the Protestant martyr and godly wife. And, while critics have been quick to light upon Ferdinand's commitment to outlandish theatrics, few have noted that Bosola makes an equal investment in performance even as he resists Ferdinand's antics.[6] Ferdinand openly mocks the performance of salvation with bizarre spectacles designed to bring his sister to desolation and despair (in effect, he out-Frankfords Frankford); Bosola, meanwhile, insists the show must go on, stepping in where Ferdinand fails as spiritual director.

Bosola understands that the performance of salvation is a commemorative script rather than a true documentation of performance; it relies for its efficacy on his skills at stage-managing outcomes and (re)interpreting the Duchess's behaviour for her audiences present and to come. From the first moments of her incarceration, he works relentlessly to cast the Duchess as a faithful stoic rather than as a woman bound by rage, frustration, and loss. Act Four opens with his description of her bearing 'so noble / As gives a majesty to adversity'; her tears take the place of her smiles to shape her 'loveliness' of old; she 'muse[s]' in 'silence' for hours on end (4.1.5–10). Shortly after this speech the Duchess shows us the nuances of her prison persona, cursing the stars and repeatedly exploiting the disavowed link between the destruction of her female body and the production of patriarchal authority. Nevertheless, after her exit in 4.1 Bosola reframes her rage against salvation's machine as nothing more than her too-passionate apprehension of 'Those pleasures she's kept from' (4.1.15): 'Send her a penitential garment to put on / Next to her delicate skin, and furnish her / With beads and prayer-books' (4.1.119–21), he tells Ferdinand, envisioning both costume and props for her role in the martyrology he is already writing. These early interpretations set the tone for Bosola's construction of the Duchess throughout Acts Four and Five and presage his Act Four-closing eulogy, in which he deploys the image of her 'sacred innocence' to fashion, tearfully, his own persona as the play's avenging hero (4.2.354). Bosola's idealizing script bookends Act Four and frames our every encounter with the Duchess within it: his words are meant to shape our expectations of her testimony and guide our reception of her acts.

Ferdinand's far more spectacular performance economy, meanwhile, relies on an inherently unstable sensory apprehension rather than on stabilizing narrative contemplation: eyes that misperceive in 'art' the hung bodies of a husband and child (4.1.111); a hand that touches what it believes to be a dead husband's hand; ears that take in the piercing cries of the mad. This appeal to the sensory returns the volatility of the body's grief to the space of Bosola's desperate attempts to write the Duchess's raging body out of his story of patient, noble suffering. It also raises the spectre of the *performing* body – the complexities of embodied perception and imagination; the communicative power of the sensory – on whose disappearance Bosola's text depends. When the (co-operative) Protestant female martyr stages her own bodily forgetting in order to gain access to history, she abdicates more than the idea of body: she also changes her relationship to sensory experience and to the power embodied knowledge wielded in early modern public space, especially among wives, widows, and young women (see Gowing, 2003). On the pyre, hands no longer signal the means of her corporeal violation, nor do they stand for the sometimes comforting, sometimes dangerous physical connectedness that shaped early modern life for women, nor – ironically – do they stand as harbingers of theatre's gestural economy, as Lavinia's missing limbs so uncannily do. The martyr claps as the flames rise (Knott, 1993, p. 9), signalling only her distance from her own body and from the physical associations that mark her body's being-in-the-world-with-others.

The Duchess, following Ferdinand more than Bosola, organizes her recusant performance practice around her resistance to this very protocol of specifically sensory abdication. She insists not only on speaking *of* her embodied experience of pain, but also *through* her embodied experience of pain, transforming her body's sensory cognition into an essential constituent of her speech. The Duchess's somatic revisions centre around Ferdinand's gift of the dead man's hand (a waxwork model meant to represent Antonio's severed hand) near the top of 4.1. The hand is a deliberate reference to the Duchess's secret wedding ceremony early in the play: 'Here's a hand / To which you have vowed much love; the ring upon't / You gave', Ferdinand tells her (4.1.43–5). The hand thus invokes the power her acts of speech and touch once carried to affect material change in her world on her own terms. But the hand also stages the body outside itself, transformed into a holy relic, separated from its power to act and relegated to the realm of hermeneutics and hero-worship, prefiguring Bosola's attempts to shift the Duchess's experience of her suffering into an abstracted spirituality. '[N]ow you know directly they are dead', he tells her after Ferdinand leaves; 'Hereafter you may wisely cease to grieve' (4.1.56–9). Refusing Bosola's prohibition against grieving, the Duchess then hearkens back to older rituals of female mourning, rituals considered particularly 'volatile' by post-Reformation English culture (Phillippy, 2002, p. 48) and, of course, totally inappropriate for a female martyr.[7] She insists that she be bound to her husband's 'lifeless trunk' (4.1.68); she invokes Portia, who committed suicide upon the death of her husband by swallowing hot coals (l. 72–4). While this kind of passionate suffering, as Christina Luckyj argues, need not be out of keeping with stoic endurance (2002, p. 136) or Christian faith, the Duchess's keening rage is more troublesome than reassuring to Bosola because she insists specifically on its somatic value. Her body's points of connection to others – arms, hands, mouth – become sites not for abstracted and isolating testimony but for the physical manifestation of loss, loss as a tactile experience (bodies roped to one another; the sear of hot coals on cheeks and tongue), loss that is made more intense because it is bound up with the memory of sensory pleasure. Bosola labels the Duchess's contraband grief 'despair' (l. 74) precisely because she will not turn her grieving body against itself, will not help him cleanse it and thus make it exemplary. For the Duchess, by contrast, to 'despair' means to place faith in the semiotic power of her skin to help make sense of loss.

Recent work on early modern sense perception by Elizabeth Harvey (2003a; 2003b; 2003c), Laura Gowing (2003), Margaret Healy (2003), and Carla Mazzio (2003) focuses on the fundamental ambivalence of touch in the period. Harvey argues that touch was associated with 'authoritative scientific, medical, and even religious knowledge' (2003a, p. 1) but also with desire, disease, and the pornographic; above all, it provoked anxiety precisely because it could not be localized in a single place in the body, or yoked to a single form of knowing. Touching was a physical act, but it also coded psychic and spiritual experience. It was grounded in the body, but it also let the body travel, mark other bodies; it acted as a metaphor for the body's interiority, its susceptibility to contamination by others, but it also simultaneously signalled 'the possibility of contact with divinity' (ibid., p. 15). Touch was 'everywhere and nowhere', feared and loathed but also craved, complicating relationships between specific senses and their corresponding bodily organs, between body and affect, body and other bodies, God and self, and, as Harvey provocatively asserts, 'drama and audience' (ibid., p. 20). At the theatre, 'To hear was to be touched', suggesting the un-nerving possibility that touch might be 'a condition of emotional receptivity, of allowing one's self to be "entered" by simply being curious' (Mazzio, 2003, p. 181; see also Bloom, 2007, pp. 111–59). Early modern touch was dangerous precisely because it risked new worlds, new configurations of body and skin: one's touch demanded another's witness in turn, made and unmade the self as it did the other. Touch transformed the body, as at the theatre, into an echo, leaving bits of that body in others, exposing the trace of the somatic within the godly, corporeal violence (the dangers of tactility) within the sanctity of the divine. Touch also signalled a broad and nuanced web of embodied knowledge, the performer's repertoire. Site of dynamic and often conflicting meanings, touch marked and shaped early modern intercorporeality even as early modern patriarchy attempted to empty certain bodies (and connections between bodies) of cultural meaning.

All of these sometimes contradictory qualities of tactility the Duchess marshals as she returns the ghost of Ferdinand's dismembered hand, queerly, to the all-important scene of her execution. Hands were early modern culture's favourite metonyms for touch, but they were also touch's master, harnessing the diffuse, web-like power of the skin organ and channelling it into a single 'signifier of domination and reason' (Harvey, 2003c, p. 88). Ferdinand seals his power over the Duchess through Act Four with his gift of the hand, but in her final moments the Duchess challenges this power by exposing the dynamic and uncertain experience of her body – in relation both to the other bodies in the room and to God's divine body – buried by the hand's claims to know and control every inch of her soma. 'Pull, and pull strongly, for your able strength / Must pull down heaven upon me', the Duchess commands her executioners (4.2.229–30) as Ferdinand's hand becomes, via her new metaphor, their hands, their strong arms, their entire bodies, directly implicated – in heaven's name – in another body's destruction. Just as Frankford's curse leaves the violent trace of conduct language in Anne's decaying flesh, the Duchess infects the script of the martyr's salvation with the force of her about-to-be-ravaged body and the bodies of those who will ravage her; in the process she turns a moment meant to enact her worldly effacement into a moment of astonishing brutality, rendering the scene of her martyr's ascension quite stunningly insecure. Although she kneels, her body moulded into the conventional pose of the supplicant, and promises to enter heaven (as princes should do) on her knees, the physical force of her metaphor crashes her final prayer, ruining the tableau of the godly woman's patient acquiescence. If to ascend is to lose the signifying power of her flesh, blood, and skin, heaven will have to come and get her.

The Duchess's command to pull heaven down upon her is the second of two occasions for the hand's spectral return in this scene. 'Bestow [my body] upon my women, will you?' she asks her executioners first (4.2.228), conjuring in counterpoint to martyrdom's stunning violence the far more mundane image of her household servants and their physical labour. While it may be tempting to read this first command as the Duchess's final memory of worldly comforts in a 'feminine' space (Haber, 1997, p. 149), the reality of her relationship with her servants is more complex. Cariola is a trusted member of her household and family, but the other women in the Duchess's entourage are not individuated. Gowing argues that women's touch was neither implicitly supportive nor simply benign in early modern England because women were the agents through which other women's places in the cultural hierarchy were policed. Spaces contemporary scholars tend to idealize as 'feminine' – such as the birth room or the shared bed – were in fact fraught with deep insecurity and could even occasion physical violence between women (2003, esp. chs 2 and 5). In Webster's text we see the Duchess in direct conversation with only one other female servant than Cariola, and only once: in 2.1 she commands an unnamed waiting woman to mend her ruff but quickly grows frustrated, chiding and insulting her (2.1.117–21). This brief moment offers a glimpse of a by no means idyllic web of female relations in the Duchess's house, and suggests that the laying-on of hands implied by her wish to be given in death to her women might be a quite complicated affair.

These two separate invocations – the long arms of heaven that reach down to strangle the Duchess, though they promise salvation; the waiting arms of her waiting women, which may or may not promise the care born of devotion – conjure the force of the Duchess's raging, vulnerable body as an independent site and source of meaning in the very moment of its requisite elision. They also mark touch as the primary conduit through which the Duchess contaminates Bosola's sterilizing performance of salvation, penetrates with conflicted affect the oppressively consistent text of his martyrology. The Duchess's performance of despair plants the somatic trace of 'godly' violence – violence rendered in God's name; violence disappeared by the invocation of God's peace – within the space of Bosola's sanctifying eulogy, but it also leaves another key remainder. In the space where her body kneels to die, the Duchess remembers the web of somatic kinships, the mundane yet risk-laden social performances that define women's bodies as those that touch and are touched (are dangerously susceptible to a host of different kinds of touch, from routine bodily surveillance to physical violence) in post-Reformation culture (Gowing, 2003, pp. 16, 53; see also pp. 206, 208). These corporeal echoes form a defining component of the Duchess's representation in Act Four, long before her ethereal reappearance as disembodied echo in 5.3. Harder to hear and yet semiotically richer, these earlier echoes offer tremendous promise to feminist theatre-makers and theatregoers as a precedent for, and a performative revision of, that later, domesticated voice. As I shift now to consider *The Duchess of Malfi* in recent performance, I want to ask: can the spectral return of the Duchess's repertoire – the trace of her performance of a fully embodied, sense-rich despair – relocate, *dis*locate, her audience in Act Five, stop us from falling 'in awe' before the 'phantasmagoric' spectacle (Simon, 2005, p. 144) of Bosola's heroic conversion, Antonio's warning echo, the Duchess's body folded so neatly into text?

Ghosts in the machine

In Peter Hinton's punk/goth *Duchess of Malfi* for the Stratford Shakespeare Festival (2006), Antonio (Shane Carty) and Delio (Steve Cumyn) arrived on the Tom Patterson stage at the top of Act Five, scene three carrying carved tin lanterns. The scene was lit very low; the lanterns swung erratically, fracturing the entire space into dizzying pops of shadow and light. While Antonio and Delio talked of the 'best echo that you ever heard' (5.3.5), the other members of the cast appeared in clusters at each of four stage entrances, their faces heavily painted to look like skulls. They seemed to have been touched, infected, by the Duchess's trauma, absorbed it quite literally into each of their own bodies, been called to 'address-ability' and 'response-ability' by this profound physical connection with her loss (Oliver, 2001, p. 7; Simon, 2005, p. 92). And yet their still bodies and neutral expressions suggested guardedness: to witness this echo was, for them, clearly a precarious gesture, and while they would be present to the scene they would not get too close. The Tom Patterson features a runway thrust stage with spectators seated along three sides; the entrances we use to get to our seats and through which we enter and exit the auditorium at intermission also function as stage entrances. The skulls gathered there, truly among us, uncanny doubles of our watching selves. Not entirely sure how to understand what they were about to see, they waited with us and with Antonio to be touched by this famous echo.

As the skulls massed, the Duchess (Lucy Peacock) entered from upstage centre. She wore a stylized farthingale, a black veil that obscured her face, and an enormous, canoe-shaped black hat that sat crosswise on her head; all were familiar from her first entrance in 1.1, the only other time in the production that the Duchess wore these carnivalesque items together. The outlandish costume distanced her from us, and her body from her persona as echo: just as the skulls bore the weight of her displaced destruction in this scene, she appeared not as the simple return of the missed and missing body, but rather as a source and a product of performance. Trailing Peacock's Duchess was Joyce Campion, the elderly actor who greeted the audience as a 'living memento mori' (Cushman, 2006) during the show's pre-set. Campion was Peacock's ever-present shadow in this production, and during 5.3 she followed the Duchess as the Duchess followed Antonio, sweeping around the stage to deliver her lines of warning. Peacock ghosted Carty's body as Campion ghosted (had always ghosted) Peacock's, as the other ghost-like characters shadowed the Duchess, shadowed Campion, shadowed our own acts of watching.

The Duchess's echo proper was, meanwhile, like everything else in this scene both an otherworldly thing and a theatricalized deferral. Peacock's voice came not from her body but from the sound system to blend with Peter Hannan's soundscape, composed of snippets of text from the play refracted through electrified liturgical harmonies. The Duchess's lavishly costumed form slid into theatre's sound machine, became ghost not to Antonio but to the scene itself, or what Alice Rayner (2006) calls 'death's double' as 'the phenomena of theatre'. Far from staging a protective Duchess as moral exemplar, the dream of Bosola's salvation narrative, Hinton shaped the 'echo' scene instead to be the culmination of the disruptive echo chain on which he built his whole production.[8] Peacock's Duchess emerged as a palimpsest, a concatenation of bodies, sounds, and images: the jumpy lantern light; the skulls at the doors; Campion as her spectral self, her self as other; the huge black hat and veil. When she entered the sound system, she was everywhere and nowhere in the space, a collection of performance fragments that could not be localized and yet seemed to touch, to penetrate, every body on the stage (and many, no doubt, in the stalls). Hinton resisted the stultifying, monumental thrust of the echo scene by locating the Duchess instead within the quirky embodiments of the living stage: never without a shadow, a witness, the Duchess disappeared into the perversely self-conscious high melodrama of the moment, leaving the force of her displaced trauma streaked on the faces of those who stayed behind.

The first time I saw Hinton's production (on 1 June 2006), I confess I was underwhelmed by all of Act Five, and especially this scene. I found the sound tricks kitschy, the Duchess's trailing of Antonio too fond and cloying. During that first viewing I was seated high up on the stage-right side of the Patterson thrust; seeing from above (from a God's-eye view?) I felt somewhat divorced from the visceral pull of the action, inclined to dismiss the scene rather than let it challenge the limits of my expectation. I took little notice of the ghosts at the entrances, and from my perch I read the scene's manifestation of the Duchess as a simple, fairly cohesive version of her Act One self rather than as a site of bodily fragmentation and performative complication. I left the theatre unsatisfied. In order to make a different kind of sense of the scene I realized that I had to go back to the beginning, to see, as Rayner suggests, what I had seen before, to let myself become vulnerable to the 'hackles of doubt' (2006, p. xxvi) the show's many ghosts raised from its first moments. The next time I saw the show, more than two months later, my seat was downstage left, immediately beside one of the auditorium's two shared entrances; the actors on stage were slightly above me, and their harried entrances and exits lifted the hair on my arms and neck as they rushed on and off. When they gathered to watch the Duchess in 5.3 they spooked me, appearing on my right-hand side with barely any sound of their own. I felt as though I was seeing the production a second time with a different body.

My two quite unique experiences of this production led me to wonder how much where I sat mattered to how (and not just what) I saw. I have since concluded that my downstage seat was not inherently superior to my seat high above the stage; the production did not play to one more than to the other, although its eerie touches certainly resonated more clearly for me down low. What mattered, rather, was my return to the scene of my initial missed encounter(s), the rehearsal and revision of my first viewing in my second. Of course two viewings of any performance are always better than one. But why? Because I catch fresh details when I'm not busy trying to grasp the story or the production ethos, but also because a second viewing of any performance always raises the memory of my earlier watching self and insists on the loss that resonates when I realize what has already passed me by. This miss between moments of perception is part of what makes performance both pleasurable and powerful. Rayner writes: 'The appointment with theatre demands that one not only witness a missed encounter *but also act it out*' (2006, p. 28, my emphasis).

Hinton's *Malfi* depended, as I said a moment ago, on an echo chain of acting bodies who were from its first scene painted to look more skull-like than human, and who grew ever-more skull-like as time passed. They played their own characters but they also doubled, multiplied, the spectre of the Duchess's torture and death throughout, even before the events of Act Four had taken place. Audience members needed to watch carefully, and to pay close attention to detail, in order to note the subtle makeup changes on the cast and to take stock of their growing import. Although many reviewers noted the heavy, macabre paint in their overall assessment of the production's style choices, chances are good that most spectators caught only a passing glimpse of this effect until Act Five, when it became especially prominent.[9] Hinton's echo chain thus produced, for audiences, a series of missed encounters with the silenced body in trauma that the skulls simultaneously invoked and displaced. As they gathered at our sides in 5.3, we acted out together *our* collective miss as *Malfi*'s defining event. But, at the same time that it damned us as insufficient witnesses, Hinton's echo chain also articulated the 'call to witness' staged by the skulls as 'an impossible demand': the demand not to 'reduce the other to an object and fail to acknowledge the other as another, separate subject' (Rayner, 2006, pp. 28–9). (This is, of course, the demand that Bosola fails to heed, that few martyrologies ever heed.) The skulls invoked and displaced the Duchess, but they also seemed to swallow her whole. Hinton refracted in their blackened eyes not only the Duchess's suffering and resistance elided by Act Five, but also the extraordinary challenge posed by the production's call to witness that suffering and resistance in their own right, as both tangibly present and yet always-already gone.

Joyce Campion, Hinton's most important onstage witness and Peacock's most fully embodied ghost, was the primary avatar for this challenge from the outset of the production. Sitting upstage centre as audiences entered, harshly downlit and made up to look absolutely ravaged by time, Campion watched us all enter and take our seats, watched us watching her with dull but intensely focused eyes. She was, from

Figure 3 *The Duchess of Malfi*, 2006. From left: Shane Carty as Antonio Bologna, Laura Condlln as Cariola, and Lucy Peacock as the Duchess of Malfi. Photo by David Hou. Courtesy of the Stratford Shakespeare Festival Archives.

moment one, the harbinger of the Act Five skulls, Hinton's ghost of ghosts. In my brief reading of Campion in the introduction I noted several ways we might understand her work on Hinton's stage: as a projection of the Duchess's elderly body, never to be; as the embodiment of Aragon family myths, fears, lost histories; as a cipher for the many bodies in the Duchess's world left unmoored in the wake of her destruction. She was all of these things and more. In her unforgettable, stylized corporeality she marked the body of the Duchess as omnipresent yet infinitely receding, but she was also a separate body in her own right, a member of the Duchess's household, a woman whose experiences were not solely defined by her role as constant shadow. Unlike the other skulls, who appeared increasingly to mirror the Duchess's pain as a province of their own and to observe her acts in 5.3 with a self-protecting defensiveness, Campion remained from beginning to end a metatheatrical incarnation of the difficult struggle fully to acknowledge another *as* other, to let that other touch and change the self without collapsing two distinct relational identities into one. As she took on the full force of the missed and missing Duchess, Campion provocatively inhabited the tensions, pleasures, risks, and doubts that ghost acts of committed, embodied witness.

Campion's principle character within the play was the Old Lady, a low-level, unnamed serving woman in the Duchess's household. As the Duchess's aged, haggard body double Campion signalled the disappearance of a powerful woman's physical authority. As the Old Lady, however, her body was even more vulnerable: it constantly risked being obscured because of its intimate association with the Duchess's far more authoritative self. In *Common Bodies* (2003), Laura Gowing examines the precarious lives of household servants in seventeenth century England, and in particular those of the elderly women who served in private homes and in their larger communities in an effort to earn some measure of protection in their old age. Because many of these women were poor, what authority they bore as elders was compromised by their reliance on their parishes for welfare; their 'intimate labour' in the homes or at the bedsides of other women, far from being a sign of close kinship or familial obligation, was commanded by those parishes in repayment for poor relief (ibid., p. 76). At once household familiars and yet entirely forgettable and expendable, older women were crucibles for the vulnerable uncertainty that marked every female body, to greater or lesser degree, in early modern socioeconomic space.

Campion's Old Lady was an omnipresent reminder of the difficult physical and emotional circumstances under which all lower-class early modern women laboured, and in this aspect of her work she modelled the ambivalence and uncertainty that govern the Duchess's experience of her relational, social body within her household circle of 'good women' (Webster 4.2.371). In a play in which violence against women is front-and-centre, spectacular rather than spectacularly elided, the most famous offstage moment is not the Duchess's death – executed by Ferdinand's men – but her delivery of a son (in 2.2), enabled by her own women. Webster's original audience would have understood implicitly how vexing those women's experiences in the Duchess's birth room could have been. Gowing paints a difficult picture of what she calls 'childbed conflicts' in the early modern period, arguing that the birth room was a claustrophobic arena beset by 'the exercise of authority and deference, and perhaps one in which relations [between women] were peculiarly strained' (2003, p. 155). The birth room reproduced the hierarchical relations between women characteristic of broader social space: the midwife bore substantial power to order those around her and was responsible for acting as official witness to the propriety of the birth. Midwives were also endowed with the power to use torture-style tactics on single mothers to extract their babies' father's names, withholding care until the information they sought (on behalf of male parish authorities) was forthcoming (ibid., p. 160).

The Duchess's delivery is not strictly illegitimate, but it is hidden and secret and thus bears the hallmarks of an illegitimate birth. While Delio implies that the Duchess has hand-picked a midwife, likely in order to ensure her allegiance, as well as procured other female helpers (2.1.166–70), no doubt her birth room remains a fraught space, filled with women (like Cariola, 1.1.504–6) who may pledge loyalty to the Duchess but who may not be in full agreement with her actions, and who may as a consequence silently question or perhaps even openly resent her authority (see Figure 3, p. 121). Among her birth room attendants the Duchess would likely have had a number of older women; the Old Lady, caught by Bosola hurrying through the palace shortly after the Duchess falls into labour (2.2.5), is clearly among that number and may even, the play implies in her timely arrival, be acting as the Duchess's midwife. This job adds another layer of complexity to Campion's place in Hinton's production. Her role of 'honest observer' (Gowing, 2003, p. 154) in the birth room would necessarily have been charged with uncertain allegiance, pressing her at once to allay the Duchess's 'torture, pain, and fear' (Webster 2.2.68–9) while also acting as a witness for patriarchy to the propriety of the event – something that might require her in certain moments to hurt as well as help the Duchess. Telescoped through this crucial yet invisible moment in the play, Campion's relationship to the Duchess appears in no way certain, in no way unambiguously deferential or inherently loyal to the body she serves, thoroughly complicating her *other* job as its performative shadow and pre-eminent witness.

How did the potentially troubling effects of this invisible moment register on Hinton's stage? As the Duchess's ghost Campion echoed her mistress's losses on her own decaying body, but as the Old Lady she articulated those losses within a stratified cultural web in which one woman's suffering, obscured by patriarchal self-fashioning, always obscured another woman's pain or doubt in turn (Figure 4). As a result, her final allegiances were difficult to parse. This double edge allowed Campion to be a particularly charged witness at the scene of the Duchess's death. She appeared at the top of 4.2 in the role of the (male) servant who, in Webster's text, announces the parade of madmen Ferdinand has ordered; her face, in fresh makeup, was more skull-like than ever. She sat beside Peacock's Duchess downstage left throughout the mad show and, like a familiar, helped her parse the madmen and women's various roles; she left quietly, however, upon Bosola's arrival, instructing the Duchess to 'question' him (4.2.114). Wentworth's Bosola

Figure 4 *The Duchess of Malfi*, 2006. From left: Joyce Campion as the Old Lady, Laura Condlin as Cariola, Ronan Rees as the Boy, Shane Carty as Antonio Bologna, and Lucy Peacock as the Duchess of Malfi. Photo by David Hou. Courtesy of the Stratford Shakespeare Festival Archives.

wore a full-headed skull mask with moveable jaw; was Campion's pronounced makeup in this scene meant to presage the arrival of this costume, or to prefigure at last the Duchess's death? Did her 'skull' double Bosola's 'fake', made-up face, or the Duchess's real agony? With whom, and with what, did she align herself in this moment? Or was she merely a servant, easily passed from hand to hand, following orders? Campion was notably absent during the exchange between the Duchess and Bosola-as-tomb maker, as well as during the Duchess's stunningly realistic strangulation. She reappeared in the middle of Bosola's eulogy, seemingly a representative of those 'good women' to whom Bosola prepares to commit the Duchess's body (4.2.371), yet she made no gesture toward that body and Wentworth did not bestow it upon her. Campion followed him as he left with Peacock slung limp over his back; whether (and how) she would finally take the body, whether she worked for him or for her in this moment, remained unclear.

Campion's role in 4.1 was even more ambiguous. During the second performance I witnessed live, the servant who entered to wish the Duchess 'long life' delivered his blessing to Campion rather than to Peacock, implying provocatively that the Old Lady and the Duchess were one and the same, that the ghost of the Duchess was already – had always already been – in the room. As Peacock's Duchess shrieked her curses, meanwhile, she shrieked at Campion, suggesting that the latter's intimacy with the Duchess was tremendously volatile, that as the Duchess's intimate she was also, foremost, her whipping-child. (In both the archive video performance, recorded on 26 August 2006, and in the prompt script notation [p. 527], Campion is herself the servant who enters to wish the Duchess 'long life'. In this role she once again bears the brunt of the Duchess's cursing reply: the prompt script notes that Peacock delivers 'Go howl them this, and say I long to bleed' directly to Campion, then exits with her in tow.) These subtle blocking choices defamiliarized the relationships in the room and lent the Duchess's curses an uncanny impact that made them harder to recuperate as part of a normalizing martyrology. The Duchess was not simply raging at injustice and its representatives: she was speaking and reacting to one marked both as her mimetic double and as her household familiar. As she cursed the stars, she appeared less as an authentically impassioned resistor than as a performing body woven into an ineluctable web of social and theatrical relations, visibly (like only a seasoned performer can be) in two places at once.

Campion, then, was not only the Duchess's ghostly double, echo of a receding body in pain; she was also her *performance* double, a kind of onstage understudy who constantly infected the play's image of the

Duchess's authentic, singular life and good death with the trace of the other bodies that made her performances possible. Campion was the echo of performance (the body, doubled) within the trauma of the Real (the missed trauma – the Duchess's, and her own), and in this capacity she was most visible in the moment she was most conspicuously absent: the moment the Duchess was strangled. The Duchess's death, as prescribed by Webster's text, is already symbolically loaded: onstage strangling produces an effect at once shockingly physical (the body flails for air, acts involuntarily) and blatantly virtuostic (the body is, plainly, *acting*). Hinton took advantage of this paradox by staging the Duchess's strangling in direct aesthetic opposition to his other major scenes, many of which featured grandiose costumes, excessive hair and makeup, and Peter Hannan's electrified soundscape. The Duchess's death scene, by contrast, happened in pared-down costumes and featured no soundtrack of any kind; save Bosola's skull mask there was nothing on the stage at this point to indicate performative self-consciousness. Peacock knelt on a bench placed lengthwise down the middle of the stage in a strong shaft of light. Her executioners pulled the cord and her hands flew up in deformed supplication, then immediately shifted into struggle as she fought hard against her attackers, pounding and gripping the bench. The scene went on for what seemed like a very long time; the only sound on stage was that of intense physical struggle. Some patrons laughed, finding the moment perhaps too real, perhaps curiously, inappropriately *un*real. We were all, one way or another, becoming more and more uncomfortable. Then Cariola (Laura Condlln) was strangled and fell into seizure; the audience gasped.

This staging was unabashedly Real, in Lacan's sense of this fraught term: it forced me into thoroughly disquieting collision with what I am not supposed to see, what is not supposed to happen on the stage/ before my very eyes. It eschewed both carefully managed stage realism (what the laughter had anticipated?) and blatant theatrical winking, offering instead a radicalized naturalism not unlike Katie Mitchell's. It openly refused the conventions of tragic performance in one of the few moments when, according to martyrological convention, a 'good', seamless performance really counts; instead, it chose to return the female martyr body in violence to the scene of her corporeal abdication messily, improperly, pseudo-theatrically. Peacock's struggling arms, legs, and torso, unaccompanied by sound or dialogue to help us assimilate her experience, underlined with a vengeance her moments-earlier call to pull heaven down upon her; her largely unmediated body collided with the violent force embedded in the line. It collided, too, with her audience's various expectations: of how the Duchess should die, of what her death should mean, of what it should require us to see. Peacock's body was outlandishly performative in its wrenching and flailing (we all knew it was only choreography), and yet it also openly resisted recuperation, either as 'only' theatre, or as the centrepiece of martyrdom's commemorative script.

Against this bare, even anti-theatrical scene, Bosola's tears and regret, his protestations of the Duchess's goodness and his claiming of her as his spiritual guide appeared, ironically, far too theatrical. Wentworth gripped Peacock's body, bent over it, crying; his melodramatic sentimentality grated against the truly uncomfortable responses created by the strangulations just moments before, and the result for me was not a continuity of feeling but a provocative dissonance. When Campion returned towards the end of Wentworth's eulogy to sit, wordlessly, on the bench where the Duchess had moments before died, she brought with her the physical reminder – and remainder – of that dissonance. Her return slipped the memory of the Duchess's gnarled web of gendered kinship networks, of the social and affective complexity of her destroyed body-in-the-world-with-others, into the moment of that body's supposed ascension, its transformation into Act Five's icon of justice and revenge. But her return also offered spectators an alternate model of witness to that proffered by Bosola in the wake of the Duchess's loss. As the old servant who was also the ghost of a great lady, as the near-silent, always marginal observer at the back of the retinue who was also the most theatrically self-conscious figure in the show, Campion quietly but consistently reimagined the play's central moments of witness as self-aware acts of embodied empathy staged across difference. Following Lauren Wispé, David Krasner defines theatrical empathy as a practice both 'imaginal and mimetic' (Wispé, qtd in Krasner, 2006, p. 265), an act of spectatorship that recognizes the role of spectator *as a role*, as a performance act integral to the ethical completion of the theatrical contract. 'I exist in the audience as observer', writes Krasner, 'but I also exist as a body reflecting on the actions and movements of another body. [...] The synergy between us is neither one-dimensional nor one-directional, but rather a dynamic give-and-take that allows experiences and consciousnesses to communicate across the footlights' (ibid., pp. 270–1). Empathy opposes sympathy in this paradigm: the latter is 'more passive', aware of another's pain 'as something to be alleviated' (Wispé, qtd in ibid., p. 264); the former is 'a strategy for understanding' (Krasner, 2006, pp. 264–5). No mere passive receptacle of her mistress's trauma, Campion was from her first, disquieting appearance during the pre-set a palpably critical observer – of us in the audience as much as of those on stage. She took the weight and the signification of the Duchess's care-worn body into her own, but she also pushed against that weight, responding quietly to its mistakes as well as to its needs, negotiating a shared identity between herself and it. That these negotiations were uneven, often uneasy, was part of the point: the work of witness Campion modelled was awkward, imperfect, at times dangerous, at times truly ambivalent. (At the moment of Campion's return to the stage after the Duchess's death in 4.2, the prompt script marginalia reads, cryptically, 'becomes human' [p. 541].) When she and Peacock crossed the stage one last time in 5.3, the ghosts massing at the entrances became *her* ghosts, too, reminders of what she had given up of herself in order to act as the omnipresent echo of the Duchess-as-loss. From beginning to end, Campion's unassuming performance insisted that to witness is not just to sympathize with the suffering, not even just to observe the missed encounter, but to 'act' that missed encounter 'out' (Rayner, 2006, p. 28) – to stage it on our bodies, as she mirrored time and again the Duchess's imminent disappearance, but also to stage the *impact* of that absence, that miss, that loss, on our bodies, as an experience that is both another's and very much our own.

The fifth act of *The Duchess of Malfi* does not end, of course, with the Duchess's echo; it ends in a flurry of homosocial violence, spurred on by Bosola's quasi-heroic revenge scheme, the Cardinal's continued malevolence, and Ferdinand's tragicomic madness. As scholars and audience members have long observed, Act Five is perhaps most remarkable for the way it marginalizes the Duchess as it brings her story to an end. And yet I believe we do the Duchess an injustice when we suggest there is little to see of her in the play's final scenes. Is she not there, spectrally, in Ferdinand's extraordinary animalia, in his inability to witness himself as subject, as *human*, any longer? Or in Julia's forthright wooing of Bosola, and in her murder (by the Cardinal's own book) for the crime of knowing too much, of overstepping her woman's bounds? I hear the Duchess, too, in the Cardinal's late cry for 'mercy' (5.5.41), a clear echo of his sister's own (4.4.352). While I do not have the space here to offer an extended reading of Act Five's most provocative ghostings, I do want to suggest that the 'echo' scene offers but one site for feminist intervention in the play's last moments. The spectre of the Duchess's losses can and should continue to haunt the scene even after her 'echo' leaves the stage at the end of 5.3, and the final few minutes of Hinton's Stratford production indicate the rich critical potential of such directorial choices. Julia (African Canadian actor Karen Robinson) died downstage centre, tipped forward onto the bench that served as her final altar, miming Peacock's earlier death pose; Robinson remained in that tableau, 'facing' upstage, her black farthingale fanned out grotesquely behind her, for the rest of the performance. The homosocial bloodbath of 5.4 and 5.5 continued, unfazed, around her. Like Lavinia's catatonic figure in Warner's *Titus Andronicus*, Julia's body was conspicuous in its forgetting as the fighting raged on: the male characters manoeuvred the rest of the action around it, leaving audiences to parse its seeming out-of-place-ness. Then, as Delio brought Antonio's son on stage to end the play in blessed reordering, Campion appeared, once more, on the balcony; she gazed expressionlessly down at Julia and at the rest of the mess below. The lights fell – slowly, slowly – on the small boy (Luke McCarroll and Ronan Rees) placed half way between Campion and Robinson on the bench at centre stage, his thin frame bookended by the remains of (at least) two missing bodies. The spot dropped on the boy, searing his outline painfully into my eyes, but as I followed the light I could not help but notice, in my peripheral vision, the two other outlines I was being asked also to witness, as missed.

Model witness

Hinton used the provocatively doubled figure of Peacock/Campion to gesture toward the essential connection between performing and witnessing, challenging audiences to step beyond the self-serving sympathy modelled by Bosola's eulogy and take up a more ethically informed position. The potential impact of Hinton's most exciting choices, however, was diminished by the extratheatrical politics governing their reception. The Stratford Shakespeare Festival relies on repeat patronage from audience members who are typically of or near retirement age, and who often travel long distances to spend their tourist dollars on hotel accommodation, meals, and other holiday trappings as part of their theatre experience. This theatre tourism is the basis of Stratford's performance economy. By offering plays done in a grand contemporary style (what I like to call 'festival realism': believable character journeys, carefully enunciated speeches, and explicatory hand gestures performed by star-wattage stage and screen actors), the Festival delivers what many sophisticated older patrons understand to be 'quality' drama.[10] Hinton's production took risks many Stratford productions normally do not, and was predictably criticized in the press for sacrificing the play's emotional content to its self-conscious aesthetic (al-Solaylee, 2006; Hoile, n.d.; Ouzounian, 2006; Portman, 2006; Smith, n.d.). Worse, the press cried,

Hinton sapped all hope from the story, leaving spectators dry-eyed and unsure where to locate their sympathies (Coulborn, 2006; Garebian, n.d.; L. Johnson, 2006; Jordan, 2006; Portman, 2006; Scowcroft, 2006; Smith, n.d.). Only one reported being unsettled by Campion's stoic, staring eyes during the preset (A. Johnson, 2006); several, however, complained that the 'skull imagery' was 'overdone' (Jordan) and the acting not 'believable' (Smith).

While I obviously do not agree with these readings, I need to take them seriously: Stratford is one venue at which conservative reviewers tend to mirror the majority audience demographic quite precisely. If reviewers resisted Hinton's anti-realist touches while gravitating toward Bosola as the most 'compelling' and 'fascinating' character in the play (J. Kaplan, 2006; Reid, 2006), then they obviously felt no impulse to re-evaluate their own (re)viewing practices in light of Hinton's echo chain of uncanny bodies. Yearning for sympathetic portrayals that might let them 'fetch a tear' for a moving Duchess (Middleton, qtd in Webster, 1997, p. 39), they were clearly uninterested in the critical empathy of which Krasner writes, or in the feminist unsettlement for which my own reading of the production hopes. I am not troubled by this: all politically committed scholarship is to a large extent prescriptive, as are all of my production readings in this book. I am, however, troubled by something else. Reading my own reactions to the show against the large body of mostly negative reviews collected by the Stratford archives, I realized that I, too, had been dissatisfied countless times while watching. Not because Hinton and his performers had refused me my tearful due, but because, in my opinion, they had not gone far enough. Too many performances had become odd hybrids as actors attempted to blend Hinton's risky, stylized choices with the economically driven expectations of festival realism. The result was almost, but not quite, the political *tour de force* I envision above. In truth, even in 2006 and at one of the largest venues for Shakespeare performance in the world, a fully committed, fully feminist *Duchess of Malfi* was not possible.

Phyllida Lloyd's 2003 *Duchess of Malfi* for London's Royal National Theatre was in this respect a different animal.[11] RNT audiences are trained in a broad range of performance traditions, making a non-naturalist *Malfi* viable at that venue in a way it may never be in more conservative festival settings. Lloyd's *Malfi* channelled old-school Brecht: Mark Thompson designed a set of broad, metallic steps to run the length of the stage and mirror the raked Lyttelton auditorium; actors not performing often posed themselves on these risers to watch the action. Together, the onstage and offstage (i.e., fixed auditorium) seating created and enclosed the narrow playing space on which the majority of the action took place: this *Malfi* was, quite literally, a product of spectatorship and its power to control, interpret, and reconfigure the meanings of the Duchess's words and actions (see Figure 5, p. 132). But if audiences, like Bosola, wanted the payoff of a saintly Duchess over whom to weep, Lloyd's production asked, explicitly and repeatedly, what was at stake for the Duchess in playing to a hungry audience. Over the course of her performance, Janet McTeer's Duchess abandoned her role as consummate player to become her own model witness: in her posthumous Act Five turn, she transformed the onstage risers from the site of her earlier, fetishized objectification into a space that shaped theatrical witness as a practice of critically engaged encounter between self-reflective audiences and traditionalist expectations.

Key to this outcome was McTeer's changing relationship to performance convention over the course of the exhausting two-and-a-half-hour production. (Staged with no interval it made real work for audiences, just as Mitchell's dark *Woman Killed* had done.) She began, like the other characters, under the sign of slick, televisual realism. The first scene was staged as a contemporary sporting event with Ferdinand (Will Keen) as a waggish sports announcer, mike in hand. Other cast members crowded the risers behind him, reacting in unison, a mob of robotic watchers. Into this space of mass scrutiny McTeer entered, powerful, aware of her power, in absolute control. Throughout the first two acts she played the CEO, projecting an authority over her brothers and her household born directly of her ability to maintain the illusion of that authority, to master power's performative acts. As she wooed Antonio (Charles Edwards) in 1.1 she encouraged him to sit in her chair, to try her position on for size; to assert her desire she stood, towering over him. McTeer's carefully controlled performances of the Duchess's power reminded her audiences of the extent to which the Duchess of Malfi is *always* invested in performance, a master of the stage from Webster's opening scene. She owns the reality effect; when that reality effect begins to threaten rather than enable her desires she properly becomes its scourge. As McTeer's hold began to break in 3.2, her persona slipped only slightly; by Act Four, however, this momentary uncertainty had given way to full-blown theatrical resistance as the production's earlier, televisual aesthetic released its late modern underside: the mediated performance of torture.

Rather than permit 4.1 and 4.2 to naturalize this torture and write her as its sympathetic victim, McTeer's Duchess adopted a tangibly Brechtian (rather than a conventionally stoic) distance in relation to her own experience. At the top of 4.1 Bosola (Lorcan Cranitch) and McTeer were positioned downstage centre, directly in front of the onstage risers;

132 *Violence Against Women: Invisible Acts*

Figure 5 Will Keen as Ferdinand and Janet McTeer as The Duchess (with members of the company) in Phyllida Lloyd's 2003 production of *The Duchess of Malfi* for the Royal National Theatre. Photo by Ivan Kyncl. Courtesy of Alena Melichar.

Ferdinand (Keen) was at a desk upstage, near the top of the risers, listening in on the exchange on a headset. Cranitch delivered Bosola's first lines describing the Duchess's noble bearing while looking at McTeer rather than at Keen, a reminder that Bosola offers us an ideologically motivated interpretation rather than a factual description of his charge. Through the first part of 4.1 McTeer appeared alternately composed and gasping, fighting tears; she was plainly vulnerable, though not yet plainly resistive to the script of noble vulnerability Bosola imagines. Then, as she sparred with Cranitch following the unveiling of Ferdinand's waxwork bodies, her composure began to unravel, along with her earlier performance persona. Alternating between rage, instability, and apparent lucidity, she gave the lie to Bosola's praise of her 'majesty' in 'adversity' (4.1.6), and yet she remained a victim worthy – indeed, demanding – of our sympathy and admiration.

But then, on her all-important line, 'I'll go pray. No, / I'll go curse' (4.1.95–6), an incredible shift. McTeer took a moment to gather herself before delivering the second half of the line, and when it came out, it was rage incarnate, pure noise. She stripped the Duchess's vaunted strength-in-adversity of its noble bearing, its composed reality effects, transforming it into a deafening aural assault on stoic endurance. No longer did she play the part of the woman in charge of her body, or the woman slowly surrendering control; now, she stood outside/beside that character, railing against the logic of the system in which she had previously acted. From this moment until the end of the scene McTeer delivered her lines directly to the audience, standing centre and facing front, lit in a small square meant to denote her prison cell. These lines were not asides; they were invocations, provocations. She shouted at acute volume, imbuing her language with the extraordinary force of her body, just as Saskia Reeves and Hattie Morahan did in the Mitchell productions I discussed in Chapter 3. In the crucial pause between 'I'll go pray' and 'I'll go curse' McTeer reclaimed control over her body's representation and articulated the Duchess's sudden awareness of the psychic and social consequences of performing the prayer and penitence for which Bosola calls. 'I'll go curse' became an undeniably political gesture: dramatically set off against the big emotion and more conventionally realist tone of her desperation and near-madness earlier in the scene, it marked the moment when this Duchess made the clear choice to radicalize her suffering and its attempted erasure.

Act Four, scene two then shifted once more the terms of McTeer's engagement with her martyrdom. Violent hands pushed her into a wingback chair downstage centre, restrained her and drugged her with a needle as madmen took over the risers, taunting her and acting out choice bits of the movie she was forced to watch: a series of video projections on the back upstage wall chronicling the destruction of her happy family. This scene – channeling the carnival midway, infected with confusing, jarring images and incredible noise – emphasized the role of the Cardinal's blasphemous Catholic Church in the family's suffering and cast the Duchess directly as a victim of religious persecution. The soundscape was utterly surreal, as were the accompanying images, suggesting perhaps payback for the Duchess's 4.1 outburst, or perhaps that we had all been momentarily transported into her altered brain. When Cranitch's Bosola finally returned as tomb-maker, McTeer was nothing like the independently minded, self-consciously stoic, courageous woman we normally associate with this scene. Her free will and bodily movement curbed by the hallucinogen she had been fed, this Duchess was forced to converse with her executioner through her restraints and the haze of the drugs. Hers was the truly invisible scaffold performance, the story no martyrology will tell: of the woman who can only be made to conform violently, artificially, through injection and subterfuge. As

Cranitch brought in the implements of execution, a bright electric light flew in just above them, turning McTeer's face intensely white. Was this the piercing light of heaven? Or was it another tool of torture, the better to see the work, fry the victim's brain? Or was it both?

Until the Duchess's final moments, everything in the latter part of this scene confirmed what the mad scene had shown us: that the Church of the Cardinal was a heresy, and it violated the Duchess in the name of a perverted heaven. Spectators may have guessed (hoped? feared?) that the Duchess's death scene would at last make clear the distinction between false gods and true, but Lloyd and McTeer refused this payoff too. 'Pull, and pull strongly': McTeer delivered these key lines quickly, as though by rote, while crossing herself; against the astonishingly bright light they stood out starkly as lines, as something we memorize, deliver out of habit, because we have to. She heaved in air, preparing for the final push, but then commanded her executioner to wait; he froze, rope stretched taught, creating a momentary tableau with his arms set, ready to pull, on either side of McTeer's head, all thrown into relief by the bright overhead light and the reflection of the scene in the glass scrim behind them. The tableau resolved itself as McTeer kneeled, quoting the pilgrim; she crossed herself again, and folded her hands before her as in prayer.

Throughout Act Four McTeer used her hands to emphasize her resistive stance against the violence and suffering to which she had been subject. As she railed against the logic of martyrdom in 4.1, she reached up and pulled down as though yanking at heaven, clearly referencing the 4.2 line with which she would later signal the beginning of her end. In this earlier moment and elsewhere, her hands disarticulated the discourses that circumscribed her; they were physical correlatives of her rage, the performative tools by which she refused her bodily dispossession. Now folded in prayer, these same hands signalled not conformity at last but Brecht's 'not-but', the theatrical formula that places two opposing realities simultaneously on display, asking audiences to read the relationship between them, choose which to believe by asking what belief means, what consequences it holds. McTeer delivered her final call for death in a quiet, distant voice that threw the moment's violence into sharp relief against her prayerful pose. Her openly rebellious body clearly did not fit this pose, nor did her words now or earlier. Were any of these things, in this moment, truly her own? Or was the Duchess's famed performance of salvation in the face of egregious tyranny made possible only by force, by the power of the drugs, by overpowering her female body both literally and figuratively?

Cranitch's eulogy, even more so than Wentworth's, delivered the sentimental sorrow missing from McTeer's final moments, but again the contrast was grating: Juliet Herd called the eulogy a descent 'into melodrama' that provoked audience twitters. Cranitch laid McTeer out on her back, placing her right hand across her left in an attempt to recreate the Duchess in prayer, perhaps imagining what her grave monument might look like. But when McTeer awoke to call for 'mercy' she rose from this enforced pose, stood gracefully, faced her offstage audience, looked down at Cranitch, and said the word in a perfectly detached, slightly patronizing voice. Bosola claims in his eulogy to see in the Duchess's briefly opened eye a reflection of heaven's mercy (4.2.347–8), but McTeer's delivery queried, even disdained, his claim. Poised between her overtly deferential body in prayer prior to her execution and the stunning detachment of this line, Bosola's melodramatics stood out easily as no true history but just another, competing performance of cultural truth.

When McTeer rose to confront Cranitch's revivification of her body in the name of heaven's mercy, she also enacted the Duchess's central transition in this production – from the role of conforming, and then later resisting, performer to the role of witness. The lights came up slightly on the risers behind her; she moved to sit a few steps up at centre, in a low spot. Her arms in front of her, hands folded over her knees, she echoed once more her earlier prayerful pose, but this time hers was the pose of one who watches rather than kneels, one who looks ahead and around rather than just up or down. She held a version of this pose throughout Act Five as she watched the rest of the action play out below, sometimes bringing her arms down to her sides, bracing herself, at other times leaning into the action, scrutinizing closely. She was not passive but attentive; she modelled critical empathy as Krasner imagines it, as a thoughtful engagement that happened across the necessary distance between herself and the others who remained enmeshed in the thick of the story. Her body reacted but it never appropriated: she did not suffer as or for others, but scrutinized instead the source and import of their suffering. McTeer rose from the play's competing constructions of the Duchess's death in order to take up the part of the witness who had thus far been missing from the production's prominent onstage audience: the witness alive to the contradictory articulations of the Duchess's body, determined to keep the complexities of that body's story in view.[12]

In this transition from performer to witness – and, importantly, *not* in the moment of the Duchess's death proper – McTeer enacted Lloyd's greatest performative debt to Brecht. Linda Kintz (2007) argues that Brecht continues to provide an important model for ethical encounters between auditorium and stage precisely because he views the performer as a witness, and the observer as a performer. Central to Brecht's theatrical model is the gap or 'pause' he commands between acting body and character body, between actor's history and character's history, and between both and the bodies, histories and experiences of those watching. In Brecht's work, Kintz argues, 'the object is no longer an object; the subject is no longer a subject', and thus 'it is impossible to master the image', to enact the cut between self and other essential to what Oliver (2001) calls 'recognition'. Brecht unsettles, then, just as Joyce Campion unsettled me: he animates the space between 'not' and 'but', the space of the missed encounter, as a place of and for spectators, in which we may act out the implications of our engagements – intellectual, visceral, affective, historical – with the performance event. He asks us to understand watching as an act of *doing* rather than being, a call to action but also a call to *enactment*, to *act ourselves differently*. Brecht turns the missed encounter at the heart of every performance event into a place for Roger Simon's 'difficult learning': a learning that can 'provok[e] deep questions about what it means for us to understand the lives of others' (2005, pp. 102–3).

Lloyd is alive to the ethical potential of Brecht's imagined witness, but she is also alive to its rarity in much contemporary Anglo-American theatre. She modelled throughout her production not only the possibility for such witness but also, alongside it, the spectre of continued audience resistance to the self-reflective watching it demands – and she used the horrific events of the play to suggest some of the real-world consequences of that resistance. Like Hinton, Lloyd received largely negative reviews (some lukewarm, some openly hostile) for her work. Reviewers complained that her approach marginalized the play's ostensible focus on 'female courage', and that her Brechtian frame-up turned the play into 'an illustrated lecture' or 'demonstration of Webster's themes' rather than letting those themes pull audiences into the play's 'deep pit of darkness' (Billington, 2003).[13] They criticized McTeer for being too chilly as the Duchess, and several reviewers read her 4.1 vocal *tour de force* as 'silly hysteria instead of psychological collapse' (Coveney, 2003; see also Scott, n.d.; Foss, 2003; and Wolf, 2003). While taking reviewers as model audience members is a far less sure-footed gesture at a heterogeneous venue like the National (which programmes a variety of work for a large cross-section of the public at all price points) than at an expensive out-of-town festival venue like Stratford (which caters to a decidedly older demographic in its theatrical programming, its pricing, and its amenities), in their general consistency Lloyd's notices do suggest the extent to which contemporary viewers may remain tethered to a pleasurably cathartic rather than a self-consciously critical experience of the Duchess's torture and death. Lloyd raised the hackles of London's theatrical establishment precisely because her production cast doubt upon the social and political efficacy of this kind of passive, consumptive watching. As John Nathan remarked in his *Jewish Chronicle* review (2003), in Lloyd's hands, 'just when you think you are about to be moved you are too often left hanging.'

Lloyd employed two related strategies for mirroring her audience. The first involved her onstage audience as it both occupied and was remembered in its absence by Thompson's imposing metallic risers. Beginning in the opening sequence, in which cast members on the risers reacted to the unseen action on the sports field as though a contemporary English soccer mob, Lloyd framed her onstage auditorium as a potentially threatening space. During the first three acts, onstage audiences of various sizes watched the play's action along with, and directly in front of, offstage spectators, mirroring the larger Lyttelton auditorium quite perfectly as they occupied the risers' steep rake. This onstage audience challenged offstage spectators not only to see themselves seeing, but also to spot the material consequences of their opposite numbers' failure to engage. An audience of holy men marked the Cardinal's arrival for his first scene with Julia: the men walked slowly to the top of the risers, folded their hands, bent their heads, then turned toward the upstage wall, away from the stage action and away from the larger auditorium. These men embodied the blind eye, the Catholic Church's steadfast refusal to witness its own abuses, and they foreshadowed the Duchess's violation as a direct consequence of such failed witness. By contrast, during the Duchess's playful bedroom scene with Antonio and Cariola a large chunk of the cast remained on the risers, watching but not reacting; they suggested constant surveillance, the thin, fraught line between public and private space in this play, as well as the possibility of seeing the Duchess in several dimensions at once. This scene marked the last appearance of the onstage audience until Ferdinand's madmen arrived in 4.2 to taunt the Duchess as she watched her brother's video; the madmen took a shocking and perverse pleasure in observing and rehearsing her trauma, one that perhaps refracted and distorted our own similar (though consciously disavowed) urge to enjoy her pain. (If we understand the 'mad' scene as a representation of the Duchess's altered cognitive function, this effect may be heightened: the madmen may, in fact, have been inside our heads, too.) During the Duchess's interrogation Ferdinand was her sole spectator, watching from a careful distance so as to avoid direct incrimination; later, the executioners and Bosola alone watched her excruciating death, her body fighting, like Peacock's Duchess, to the last (Figure 6). Increasingly, Lloyd challenged offstage audience members to take on the vacated role of onstage watcher, to imagine ourselves onto the risers so lately and problematically animated by the cast's acts of 'dangerous seeing'. Not every audience member will have understood this challenge in the same way, of course, but few will have failed to notice how Lloyd's stage watched us, just as it watched – and then destroyed – the Duchess, provoking at least one or two 'deep questions' (Simon, 2005, p. 103) about what exactly we had come to see.

Lloyd's second strategy for reflecting (on) spectatorship was, like her first, a design choice. Directly in front of the risers, between them and the playing area, Designer Mark Thompson placed a sliding glass scrim. Both reflective and transparent, it moved in order to mirror certain elements of the onstage action; it also acted as a screen behind which

Figure 6 Janet McTeer as The Duchess and Lorcan Cranitch as Bosola in Phyllida Lloyd's 2003 production of *The Duchess of Malfi* for the Royal National Theatre. Photo by Ivan Kyncl. Courtesy of Alena Melichar.

characters could watch or listen without being seen. But the scrim, being glass, refused to allow bodies to disappear: instead, it generated body doubles, captured performer and observer within the same representational space. The scrim made palimpsests: it revealed the bodies that lay behind and beneath the body we were meant to see, showed us how that body was marked, framed, shaped by the actions and interpretations of others. And more: it also mirrored us, reflecting a portion of the larger auditorium within that same space. Both of the Duchess's audiences thus became visually and performatically implicated in her tortured embodiment, as well as in the play's attempted erasure of that embodiment. While chances are slim that a single audience member seated in one of the rows directly in line with the scrim could have seen him or herself reflected clearly, that scrim nevertheless staged auditorium space *as* playing space, wove the trace of the audience into the world of the Duchess's experience as a central participant rather than a peripheral (and invisible) observer. During Act Four the scrim trailed the Duchess relentlessly: it doubled her body, bent double; it doubled her rage; it doubled her execution. It staged the possibility of another Duchess *behind* the Duchess spectators may have longed to see; it staged the need to bear witness to a Duchess we do not immediately recognize.

* * *

Both Lloyd's and Hinton's productions articulated the failures of witness that define this play as a seventeenth-century martyrology, but both offered something of the promise it holds for contemporary feminist performance as well. Both productions put their audiences on stage: they deliberately located acts of performance and acts of witness within the same epistemological space, reminding us that we exist in the frame of performance, that our acts of spectatorship carry with them material consequences, 'response-ability' both to those on stage and to ourselves. Both productions asked us to consider what it means to watch (and to be watched by) the Duchess of Act Five in the wake of the Duchess of Act Four, to see in Act Five not Act Four's denouement but its missed encounters. Both productions resisted the transformation of the Duchess's violated body into a simple, unifying, and isolating text by foregrounding her social, affective, and performative relations with other, less visible bodies in her world (including our own spectating bodies), and by radicalizing her performance of despair with deliberately unconventional, oppositional stagings of her final scene of sacrifice. Finally, both productions unravelled the sentimental power of 5.3 by reminding their audiences that the salvific echo the Duchess is meant to become is also, foremost, an actor, a palimpsest of bodies, her melodramatic voice just another organ of the stage. In each of these productions the Duchess was everywhere and nowhere, everyone and no one. She was refracted in the skulls that massed on the edges of the Patterson stage, in Campion's aging face, in the doubles and distortions of Lloyd's glass scrim, in the apathy and cruelty of her onstage audience. And she was in us: in each of these productions, *we* became the Duchess's most profound echoes, the actors left behind, our voices imbued with the power to articulate what remains – after Bosola's eulogy, after 5.3 – unspoken.

5
The Architecture of the Act: Renovating Beatrice Joanna's Closet

> Dorothy falls and Alice falls, but into other worlds of magic and strangeness.
> Adam, Lucifer, Humpty Dumpty, and Icarus fall to less desirable ends. These figures of the construction called masculinity attempt to rise to power and fail, lose, fall from grace.
> The feminine ones drop out, fall down the hatch, use the exits.
> (Bloomer, 1993, p. 162)

This final chapter examines the in/visible act as a function of theatrical architectures. It takes Thomas Middleton and William Rowley's *The Changeling* – a play about the sexual violence of space – as its central analytical object, and it marks both the culmination of, and a departure from, the work this book has done so far. I am interested in two related problems here: the spatial dynamics of sexual violation (how a female character's locatedness on stage positions her for rape, and more), and the spatial dynamics of reception desire (how a play text or a live production of that text positions *us*, in the audience, to read a female character as sexual vixen or rape victim, sufferer or perpetrator). This last concern, in particular, represents the climax of my preoccupation throughout the preceding chapters with acts of spectatorial percepticide, the possibility of audience witness, and the relationship both hold to the ways in which we currently imagine, and might productively reimagine, the representation of violence against women in early modern performance. The in/visible act is not simply a politicized performance gesture, a making-visible of the historical invisibility of violence against women: as the Duchess of Malfi knows, it necessarily requires a parallel gesture of *politicized spectatorship*, a willingness on our part to be unsettled, to come to terms with the 'difficult knowledge' it imparts about the current shape of our viewing practices (Simon, 2005, p. 10). The risks and the potential of reception desire have ghosted every one of my readings thus far, because the politics of reception lie at the heart of the ethical questions that shape this book. What are we willing to see of a woman's violation on stage? What is it our responsibility to see? How does the shape of our watching collude with the narrative and performance structures that enable violence's effacement? How can we see beyond those structures to the ideological shape of violence's representation, the stakes of our own watching?

It was Beatrice Joanna who first prompted me to ask these questions. Hers is the story of a woman undone first by architecture, and then by us.

Rape, space, and reception desire

At the top of the *The Changeling*'s brutal Act Five, scene three, Alsemero and his confidant Jasperino exchange brief words. Jasperino suspects Beatrice Joanna (Alsemero's wife) of adultery. He invokes a 'proof': 'The prospect from the garden' (5.3.1–2). Jasperino and Alsemero seem to have caught Beatrice Joanna in the act with her 'lover' De Flores; Alsemero, now forewarned and forearmed, will shortly confront his wife and set in motion the events leading to her murder. And yet, certain though this tense exchange between friends may seem, it leaves unanswered a key question: just what, exactly, is this 'prospect' of which Jasperino speaks? 'From the garden' suggests that his and Alsemero's spying eyes were watching at some distance from their subjects, and yet Jasperino has little doubt of what they have seen: no less than Beatrice Joanna's sexual desire, handmaiden to her sexual betrayal. The possibility that the 'prospect' might have captured something else of Beatrice Joanna's experience – coercion, trauma, shame, another in a series of ongoing sexual violations at De Flores's hands – does not, indeed cannot, occur to Jasperino. If Beatrice Joanna has been observed in a compromising position with a man other than her husband, only her apparent 'sexual defection' marks the 'violently offensive' heart of this invisible scene (Burks, 1995, p. 766).

'The prospect from the garden', of all the horrid prospects offered (and denied) by this play, may seem an odd spot for a feminist intervention. It is brief, practically an afterthought, and it appears unequivocally to indict Beatrice Joanna. And yet this scene that is not a scene – that is not even narrative, really; that is no more than a perfunctory phrase balanced on the knife-edge of an angry man's assumption – forms the knot I want to worry in this final chapter. *The Changeling* is a play about views missed, prospects misread, suggestive sounds emanating from scenes tucked, tantalizingly, just off stage. It is a play in which spaces-off repeatedly mangle the distinction between sexual violence and obsessive passion, lust and trauma, calling on spectators to fill in the endless blanks vacated by theatrical image and narrative certainty. The prospect from the garden is just such a blank. It does not show us Beatrice Joanna *in flagrante*, after all; it tells us only what Jasperino *imagines* Beatrice Joanna to be doing and feeling. The prospect damns her (as she was damned at the end of 3.3, dragged off stage by De Flores to become his 'lover'; as she will be damned late in 5.3, sent off stage once more to 'rehearse' her 'scene of lust' with him) with an act that remains forever beyond our sight and hearing, and yet the space of the men's confident exchange is large enough to deliver the guilty body as if before our eyes. Faced with this prospect that is, in fact, no prospect, we have only two choices: to take Jasperino's wonders for signs, or to reframe his conjuring.

The 2006 Cheek by Jowl touring production of *The Changeling*, directed and designed by company co-founders Declan Donnellan and Nick Ormerod, chose the latter option.[1] At the top of 5.3, as Jotham Annan's Jasperino spoke of the 'prospect from the garden' to Tom Hiddleston's Alsemero, he opened the door of Alsemero's closet – a heavy, institutional-grade door, complete with glowering 'exit' sign above, built into a pillar upstage centre. Just behind the door, Beatrice Joanna (Olivia Williams) and De Flores (Will Keen) stood facing one another. Keen's left arm was raised, holding the doorframe, but otherwise the pair did not appear to touch one another. Williams faced upstage, looking at Keen, while Keen faced the audience. His expression was empty, while hers – indeed, everything about her experience at this moment – was invisible. There was no sex in this image; there was no violence, either. There was nothing to see. In fact, this tableau posed the very question that reverberates behind Jasperino's formidable accusation against Beatrice Joanna. What might he *not* be seeing?

Donnellan and Ormerod's choice to set the 'prospect from the garden' within the space of Alsemero's closet – *The Changeling*'s dead centre, both figuratively and literally – was not insignificant. The scene used the physical shape and symbolic significance of the closet to frame two ambiguously positioned bodies for a very selective, highly speculative reading of their interaction, one that resulted directly in Beatrice Joanna's final rape and murder minutes later behind that same closet's closed door. The 'prospect' thus generated not misogynist certainty, but rather provocative links between architecture (the closet's interior, at once transparent and opaque), audience interpretation (what we imagine we see within), and violence against women (which the closet both enables and elides) in the play. Simultaneously, the scene suggested that manipulating the play's built spaces in unexpected ways might have tremendous potential for a feminist staging of *The Changeling*. Alsemero's closet is a private male space, a zone of learning and self-making, and yet its contents (tests for virginity and pregnancy) imply vexing *un*certainty, an anxiety over proof rather than proof itself. The closet represents Alsemero's protected inner world, yet in Act Four Beatrice Joanna invades it, 'violate[s]' it (Jardine, 1996, p. 127), claims its knowledge for her own, momentarily shifting its cultural power from him to her. And all the while, of course, the closet is also a performance space: it is the wide main entrance and discovery space of the Phoenix Theatre (the Cockpit in Drury Lane) where the play was first performed (Orrell, 1985, p. 51), the 'backstage' of the modern thrust or proscenium. 'Exit' marks the spot: actors use the closet to come and go, slip in and out of character, cross the limen between 'fake' and 'real'. More performative phantasmagoria than anything else, the closet may just be the 'exit' of which Jennifer Bloomer speaks in my epigraph above, an escape hatch into which Beatrice Joanna might fall and claim another life.

But Alsemero's closet is also no such space, because it is also the container in which his ownership over her body and/as his private space is confirmed in one fell blow at the end of the play. Set in a fortress built to encase sexual 'secrets' (1.1.159), *The Changeling* is every inch a play about the consequences of architecture for violence against women in early modern culture, and the specific architectonics of its acts of violence require a feminist reckoning if *The Changeling* is ever to be politically viable on twenty-first-century stages. This chapter attempts such a reckoning. I begin by asking what role built space plays in the way we understand, or fail to understand, Beatrice Joanna as a victim of both sexual and physical violence. I briefly cite the history of *Changeling* reception in Britain and consider how the dimensions of performance embedded in Middleton and Rowley's text, along with the superstructures of our contemporary classical stages, shape our association with Beatrice Joanna at the moment of her rape in Act Three. I then read early modern architecture theory's disavowed debt to female sexuality and the consequences of that disavowal for Beatrice Joanna's two pivotal experiences in Alsemero's closet (in 4.1 and 5.3 respectively). In the second half of the chapter I explore the

2006 Cheek by Jowl production architecturally. I argue that Donnellan and Ormerod's staging choices effectively politicized the spaces of Beatrice Joanna's suffering while also up-ending – quite literally – the comfortable viewing positions that feed audience certainty about who this woman is, what she feels, and what happens to her. The production invested heavily in the relationship between place, power, and violence integral to the play's fictional world, but it also explored the relationship between place and power that obtains *in performance* between actor, character, and audience. Its defiant spatial politic envisions what I want to call an architecture of feminist performance for this otherwise patently anti-feminist play.

My arguments throughout this chapter pivot on my central claim that *The Changeling*'s internal spatial dynamics possess a formidable, though often invisible, power to shape the way an audience sees and to manipulate what an audience wants to see. I am interested in the ways in which the play's performance spaces position bodies – character bodies, acting bodies, but also spectating bodies – for a specific view, framing reception desire architecturally as well as narratively. By 'performance space' I mean: a combination of the conception of space and place within the play's fictional world; the spaces created in the theatre on a case-by-case basis in order to play these conceptual locations in production; and the several ways in which a production frames its audiences in time and space (Tompkins, 2006, p. 3; see also McAuley, 1999). Following the influential work of Henri Lefebvre, Edward Soja, Edward Casey, and others in phenomenology and human geography, theatre scholars now understand that space is a product of movement, a performative human creation (Casey, 1997, p. 229; Lutterbie, 2001, p. 126). How we use space every day shapes cultural meaning: the structures of our homes enable sex and gender norms; the structures of our public spaces dictate race and class norms, marking some bodies as well-placed and others as out of place, even threatening. Our movements through these spaces may work to reinforce such norms as natural and inevitable, or they may resist the codes of placement, map a renegade terrain. Stage spaces and the specific performance locales to which they give rise in each new production are both mimetic representations of extratheatrical place as well as zones of cultural power in their own right; they possess the world-shaping powers of 'real' space but also the potential to let us stand momentarily beside (if not outside) that space and observe its operations. And, as we stand both within and beside our place-worlds at the theatre, we need to account for relations among stage space, auditorium space, and the other spaces within a given theatre building (see

Carlson, 1989). How we move through these 'other' theatrical spaces before and after the show and at the interval matters as much as the apparently fixed positions of our assigned seats to the meanings we take away. Beatrice Joanna, then, is made both in the several positions from which we watch her, and in the diegetic spaces through which she is framed for our consumption. By queer(y)ing the shape of *The Changeling*'s performance spaces, directors and designers might allow performers as well as spectators to inhabit different paradigms, to act and to see Beatrice Joanna's story from literally new places.

Looking at Beatrice Joanna

The Changeling stands last in this book's critical trajectory because it telescopes with special urgency the crime of violence's effacement, and yet also demonstrates with particular elegance how easily we might use that effacement against itself in order to reposition one of the most thoroughly condemned women on the early modern stage. This is a play in which the central female protagonist's status as a victim of sexual violence (indeed, of any violence at all) is wholly uncertain, and is thus largely ours, as an audience, to determine. What will we do with her? And what might be the consequences of our 'doing' for women, not unlike Beatrice Joanna, whose violations in our own time and place are routinely called into question by their sexual pasts, their social or racial status, their criminal records? In a recent essay about *The Changeling* on the London stage (2004), Roberta Barker and David Nicol trace the play's critical reception over the last half century; they conclude that a general continuity in audience expectation, driven by a consistent critical response to major productions, has played a substantial role in manufacturing Beatrice Joanna's consent to De Flores in Act Three, scene three, turning his rape of her into a modern, even universal, scene of romance. Reviewers, they demonstrate, tend to 'explain' the play's main plot by reading Beatrice Joanna either as a woman 'destroyed by her own sin', and thus justly punished rather than unjustly raped, or as a woman on the cusp of a sexual awakening that will be brought home when her repressed attraction for the supposedly loathed De Flores finally surfaces. Barker and Nicol's research further shows how reviews of each new production build directly upon the critical consensus generated by the last, producing 'accepted interpretations' of the play that continue to shape new audiences' expectations of a specific kind of sexually voracious anti-heroine. Again and again, the London reviewers dwell on the moment when Beatrice Joanna tumbles from sexual innocence into De Flores's arms as the wellspring of the play's urbane modernity (its pre-Freudian insight); in turn they generate audience desire for the 'pleasure' that comes with the anticipation of watching Beatrice Joanna fall into her sexual destiny.[2] The productions and reviews Barker and Nicol examine all come from major London revivals of the play, produced in big-budget, large-market West End theatres. The 2006 Cheek by Jowl production I read in this chapter is also such a revival: it toured to large and well-known festival venues in Europe, and ran for a month at London's Barbican Centre as the inaugural production in Cheek by Jowl's formal three-year residency there. Barker and Nicol amply show how reviewers in the mainstream British press have colluded to produce a specific Beatrice Joanna for local London audiences (and, by way of the ubiquity of West End performance, broader British and select overseas tourist audiences as well). I want now to back their story up a bit, and to think about how the spatial organization of stage and auditorium in the large modern theatres that routinely host *The Changeling* may work in tandem with these reviewing trends in order to reinforce the impression that audiences know Beatrice Joanna's unconscious desires in an intimate way she cannot personally access. While I realize this argument risks rehearsing a long-standing prejudice against such theatre spaces and in favour of black box or laboratory spaces that appear to promote greater audience-performer interaction (see, for example, Wiles, 2003, pp. 240–66), I aim not to pass judgment on the relative merits of festival-style main spaces and their (often marginalized) studio siblings. Instead, I want to explore the unanticipated consequences of what David Wiles calls the 'obvious correlation between monumental buildings and classic plays' (ibid., p. 265). Do the hierarchies of space in 'monumental' theatre buildings – in which a typically vast, comfortable, raked auditorium, connoting privilege if not outright luxury, looks down on the stage, while the stage disguises the backstage spaces of its making – and the circulation of social and cultural capital they permit have something to do with the way Beatrice Joanna is received and (re)produced as an anti-heroine who either deserves or secretly wants what she gets? If so, what difference would it make to renovate those theatres' inbred spatial relations, to uncouple reception desire from architectures of production and reception? If, as Steven Mullaney remarks in relation to the theatrical topography of early modern London, the place of performance 'is endowed with meaning, and with power' (1988, p. 17), can contemporary audiences accustomed to assuming a position of invisible moral and juridical authority over a character like Beatrice

Joanna be made to see themselves *sitting*, to recognize their seats of power as integral to the broader landscape of the play in production?

Beatrice Joanna is unlike any of the violated heroines I have discussed so far. She is neither an innocent victim of outrageous and condemnable torture (Lavinia; the Duchess) nor is she guilt-ridden and repentant for the crimes she commits (Anne Frankford). Like the Duchess, she insists on her own sexual choice (Alsemero, not Alonzo de Piraquo), but unlike the Duchess, she engages in criminal activity in order to secure that choice, and she makes the very naïve decision to hire an untrustworthy servant (De Flores) as hit-man along the way. Like Lavinia she is raped off stage between acts, but unlike Lavinia her status as raped cannot be guaranteed by her chastity: she has behaved, as De Flores accurately points out, immodestly (3.3.126) in contracting him to murder the man to whom her father has matched her, and therefore has compromised her physical virginity with unchaste behaviour (Bamford, 2000, p. 17). If she can be called neither a virgin nor a modest woman, she can only with difficulty, if at all, be recognized as a rape victim.[3] More importantly, Beatrice Joanna's audiences are not granted privileged access to the knowledge that escapes Lavinia's befuddled relatives in Act Three of *Titus Andronicus*: we do not see malicious aggressors set upon a vulnerable victim, nor does Beatrice Joanna, as 'A woman dipped in blood' (3.3.126), have access to accepted codes of chaste resistance. Unable to appeal either to her innocence (ll. 83–7) or to her honour (ll. 120–6), she can only offer De Flores money, 'gold and jewels', 'all the wealth' she has (l. 156), before finally accepting his argument that the 'murder' for which she has contracted him will inevitably be followed by 'more sins' (l. 163). De Flores grabs the last word, characterizing her as 'blush[ing]' and 'yielding' (ll. 166–70) as he escorts her off stage, leaving audiences genuinely uncertain as to whether or not Beatrice Joanna has, in fact, finally consented.

Of course, plenty in 3.3 heralds rape. De Flores is coercive, at times openly threatening, and Beatrice Joanna expresses her resistance as a mixture of shock, confusion, and terror. The scene's final moments even include an explicit rhetorical reference to rape: De Flores paints Beatrice Joanna as a panting turtle dove, echoing the Duke's forced sex with Bianca in Middleton's earlier *Women Beware Women* (Chakravorty, 1996, p. 157; Bawcutt, 1998, p. 12; Malcolmson, 2001, p. 156; Haber, 2003, p. 83). This line, however, also references marriage: both Judith Haber (2003, p. 80) and Lisa Hopkins (1997, pp. 154–5) note an echo to Ben Jonson's wedding masque for Frances Howard, the real-world model for Beatrice Joanna. Beatrice Joanna's failed resistance and

De Flores's *double-entendre* combine to end this crucial scene with deliberate confusion *despite* the apparent imminence of sexual violence. Is Beatrice Joanna about to become a rape victim? Is she, having already forfeited her chastity in her earlier dealings with De Flores, a sexual defector regardless of her protests (Burks, 1995, p. 766)? Or does she indeed protest too much?

These are the questions that chase Beatrice Joanna and De Flores off stage at the end of Act Three, and into what has become the conventional interval for the play in modern performance.[4] The act (of sex? violence?) vacates the playing space: the actors playing Beatrice Joanna and De Flores disappear backstage, to the dressing rooms, to the places that support performance's production; meanwhile, the house lights come up and the audience stirs for the bathrooms, for the drinks kiosk, reads the programme. While the actors prepare for the second half and the characters get on with whatever it is they are doing, the question of Beatrice Joanna's status in that doing, the politics surrounding her violation and the problems of belief it raises, rebound upon us. Sex or rape, whatever it might be, happens in the non-space of the interval, in an elsewhere beyond the reach of performance dictated exclusively by the needs of the audience. As audience members perform the privileged rituals of the modern classical spectator – chatting with friends, eating ice-cream, watching one another, waiting to use the bathroom – its meanings appear and disappear among us, in the dress circle and the stalls, in our entre-act chatter, our debate over what we've just seen. Or, quite possibly, in our utter indifference to it.

The Changeling stages Beatrice Joanna's rape by De Flores as a hole at the centre of the text and its performance. The play would have encouraged early modern spectators, as Deborah Burks has convincingly argued, to observe in Beatrice Joanna's altercation with De Flores the naïvety and untrustworthiness of women, the 'traitorous sexuality to which the "consent and complicity" statutes' – changes in rape law during Elizabeth's reign – 'were a reaction' (1995, pp. 768, 776). For a modern audience, though, to whom the nuances of early modern rape law are foreign, 3.3 invites adjudication based on normative assumptions circulating in contemporary Western culture about rape victimhood – about who may legitimately claim to have suffered sexual violation, and about for whom 'no' probably means 'yes'. In something of a disruption of the modern theatrical intermission as personal, individual time, as a chance for us to reclaim the theatre's spaces for ourselves, *The Changeling*'s charged interval asks us to find a way to reinvest in the performance moment just passed and to make a decision. Will we bear witness to the complexities of Beatrice Joanna's victimhood and her (inevitably) failed resistance? Or will we retreat to a belief in her guilt, her hidden sexual hunger, and thereby assuage the uncomfortable questions 3.3 provokes about who qualifies for suffering? For these 15 or so minutes, before the production we are watching has a chance fully to weigh in with an opinion, the choice, the responsibility to choose, is ours alone.

The non-space into which Beatrice Joanna and De Flores disappear at the end of 3.3 returns, much later, in Jasperino's invisible prospect from the garden; it then returns one final time behind Alsemero's closet door as Beatrice Joanna dies. These non-spaces threaten Beatrice Joanna, obscure her side of the story, but in their palpable, pricking absence they are also zones of tremendous potential, of 'alternate ordering' (Heatherington, qtd in Tompkins, 2006, p. 95) that 'may coexist with, interact with, and resist the "real", mimetic one' (Tompkins, 2006, p. 95). Joanne Tompkins, following Kevin Hetherington rather than Michel Foucault, calls such spaces heterotopias: they '[unsettle] the flow of meaning' produced by more straightforward spatial relationships in the theatre (Hetherington, qtd in ibid.), creating the 'uncanny' sensation 'of being in place and "out of place" at precisely the same time' (Tompkins, 2006, pp. 12, 95; see also Chaudhuri, 1995, p. 138). Heterotopia in this formulation is less non-space than meta-space: the feelings of displacement it generates invite us to question the relationship between emplacement and epistemology, to consider how where we think we are, relative to our others, determines what we think we know about ourselves and those others.

The 'prospect from the garden', the closet, and the interval that contains the rape that is a hole in this text, in its stage, in both its playing and its watching are for me potential heterotopias precisely because we cannot fully access them. They haunt the edges of both theatrical possibility and audience perception, mark the spot where performance comes to an end and audience anxieties, desires, expectations, and assumptions become the primary source of theatrical meaning. They can, of course, easily be – have too often been – played to seduce audiences, played as constricting, stifling spaces that trap a doubtless naughty girl. How, then, do we play them to unsettle, play their heterotopic potential? By exploiting their inherent self-consciousness, their tangible nothingness, their playful, performative status. By locating them both within the castle's fortressed walls *and* within the malleable and multiple spaces represented by theatre's elastic fourth wall. This evocative dislocation allows for a more expansive metatheatrical return than the one Lavinia, in her failure as a performer, inadvertently produces during *Titus Andronicus*'s hysterical third act. Lavinia shows us the cruelties of rape's performance demands; Beatrice Joanna shows us the fraught, often cruel vicissitudes of the return's spectatorial requirements. A heterotopic staging of *The Changeling* literally positions its audiences to discover, in mind but also in body, how the play's spaces frame our watching and limit our acts of witness, producing again and again a specific version of Beatrice Joanna – sexed but not violated – as a universal performance norm.

Crimes of architecture

> I have seen an incredible weight, a whole mass of stone, disturbed by a single root.
> (Alberti, 1988, p. 73)

> 'Twas in the temple where I first beheld her,
> And now again the same [...]
> (*Changeling* 1.1.1–2)

Space is always double in *The Changeling* because Beatrice Joanna, she of two names, is always double in *The Changeling* (Liebler, 2003, p. 375). Middleton and Rowley structure the play around an uneasy equivalence between its heroine, her chastity, and the spaces designed to contain both her and it. She is Vermandero's 'castle' but also its 'best entertainment' (1.1.194–5); she is the statuesque figure that adorns the 'temple', the 'citadel', and its 'promonts' tops', but she is also the keeper of its 'secrets' (ll. 157–9). Beatrice Joanna's body, her beauty, and her presumed innocence are all architectural objects. The walls that surround her are the walls she decorates; they define who and what she is, but she also defines them, shapes their symbolic potential, and enacts their structural limits. The very spaces that are meant to keep her safely bound ultimately, and fearfully, prove most available to her manipulation: Alsemero's closet (4.1) and his bedchamber (5.1); her maid's bedchamber (5.1); the corridors and inner passages where she and De Flores meet, talk, and plot throughout the second half of the play. Lena Cowen Orlin notes that the conceit of a man's home as his castle became proverbial in the sixteenth century, just as the 'era' of 'the old architectural form of the feudal lord's fortress' ended (1994, p. 2). The 'castle metaphor' fast became emblematic of 'a vision of the secure and uncontested authority and "repose" of the individual English householder' (ibid., p. 5), and yet the early modern period was also caught up in the anxiety that the house-as-castle was 'too *un*like the fortress', 'too open, penetrable by and hospitable to any number of disorderly and masterless men, analogized implicitly with the female body in its stubborn resistance of male control' (ibid., p. 8). Vermandero's castle-*cum*-fortress, perched between old and new domestic formations, proves vulnerable to 'an interloper and adventurer' (Jardine, 1996, p. 122), if not a 'masterless' man, in Alsemero, but its walls are ultimately toppled from within.

By equating his daughter's chastity with her physical containment and yet al*so* imagining her as the most compelling emblem of his castle's mysteries (its 'best entertainment' as well as its 'secrets'), Vermandero runs headlong into the paradox that attends all early modern domestic building. Beatrice Joanna is the body in that building: she figures simultaneously as its content and its structure. She appears enclosed within Vermandero's walls, but also equivalent to them; indeed, she appears *as the measure* of their beauty as well as their strength – but only as long as her sexuality remains properly cloistered. Early modern great houses were 'a nested system of enclosed spaces' (Wigley, 1992, p. 340); women were confined 'at the greatest distance from the outside world, while men [were] exposed to that outside' (ibid., p. 332; see also Howe, 2003, p. 68). Amanda Flather cautions that early modern men and women likely read the prescriptive literature advising the cloistering of wives with a grain of salt, especially in middle-class homes (2007, pp. 31–2), but Alice Friedman's analysis of the history of Wollaton Hall in Nottinghamshire demonstrates the extent to which architecture could shrink a woman's world regardless of her social class. In great houses servant women were somewhat freer in their movement than the mistress, while the latter, her daughters, and attendants had access to only a limited number of rooms on a limited number of floors (1989, pp. 49–50). The wife was the household manager, to be sure, and as such was endowed with the keys and a fair amount of household power (Flather, 2007, pp. 34, 46), but good household management, especially among the nobility, meant less, not more, household mobility.

Shaping space according to the ideological dictates of the culture, early modern architecture operated as 'a system of surveillance', ordering bodies and producing gendered hierarchies by framing walls, doors and windows as 'way[s] of looking' (Wigley, 1992, pp. 339, 341; Hills, 2003, pp. 8–9). And yet, as Mark Wigley argues, despite appearances to the contrary, early modern men were spatially fixed while women were mobile. A man moved in and through the public realm beyond the household but remained forcefully attached to, defined by, that household as its head (we might think here of Master Frankford); a woman, meanwhile, travelled between houses as she advanced from daughter to wife to widow, perhaps to elderly dependent (we might think here of the Duchess's Old Lady). As an enclosure the house warded off the possibility of any transgression implied by women's constitutive mobility, but containing the *donna mobile* also meant transforming that house, even as it was increasingly understood as a realm of feminine authority (Friedman, 1989, p. 180), into the image of masculine power and control (Wigley, 1992, p. 336). This is the paradox within which *The Changeling* labours: house is defined by and shaped around woman, but it also stands, awkwardly and unevenly, for 'woman-as-housed' (ibid., p. 337).

Renaissance architecture took its primary cues from the legacy of Vitruvius (Payne, 1999, pp. 70–1), the Augustan builder and geometer who based his own theory of structural decorum on a body that has come to be known as Vitruvian man – an idealized male figure, six feet tall and barrel-chested, who, when laid on his back with arms and legs outstretched, produces the dimensions of a perfectly squared circle. Indra Kagis McEwen argues that Vitruvian man is Augustus himself, and through him the Imperial Roman project (2003, p. 12), but a feminist reading of Vitruvian man suggests a less immediately visible referent:

> [...] in the human body the central point is naturally the navel. For if a man be placed flat on his back, with his hands and feet extended, and a pair of compasses centred at his navel, the fingers and toes of his two hands and feet will touch the circumference of a circle described therefrom.
>
> (Vitruvius, 1960, p. 73)

The navel is ground zero of Vitruvius's classical system, but it is also the locus of a primal relation between mother and child. And yet there is no mother, no woman whatsoever in Western architecture's classical origin story. Vitruvian theory implies a constitutive link between the ideal male body and the harmonious representation of his power in built space, but only by first erasing the sexualized female body (the missing body of Vitruvian man's missing mother) from the scene. Her absented body then returns, uneasily, as the object rather than the subject of classical form and control. Or, as Wigley puts it, 'classical architecture theory dictates that the building should have the proportions of the body of a man, but the actual body that is being composed, the material being shaped, is a woman' (1992, p. 357).[5]

During the classical revival of the fifteenth and sixteenth centuries, the uncertain place of the female body – and in particular of female sexuality – within Vitruvius's architectural model circulated as a pervasive anxiety over the status of *ornament* in proper building practice. Vitruvius's Renaissance interpreters – including Leon Battista Alberti, Sebastiano Serlio, and Andrea Palladio – all argue for the central place of ornament, symbolized by the all-important columnar orders, but they also define their theories of architecture by setting core limits on ornament. Ornament enables all that is strong and beautiful in a building (Alberti, 1965, p. 113), but it also symbolizes the possibility of unwanted excess, of a self-conscious materiality that goes one step too far and risks spoiling the whole structure (Ingraham, 1998, p. 102). For Alberti, ornament's value lies in a builder's attention to proportion, which will result in a building's overall 'congruity' (1965, pp. 194–5):

> The chief and first Ornament of any Thing is to be free from all Improprieties. It will therefore be a just and proper Compartition, if it is neither confused nor interrupted, neither too rambling nor composed of unsuitable Parts, and if the Members be neither too many nor too few, neither too small nor too large, not mis-matched nor unsightly, nor as it were separate and divided from the Rest of the Body: But every Thing so disposed according to Nature and Convenience, and the Uses for which the Structure is intended, with such Order, Number, Size, Situation and Form, that we may be satisfied there is nothing throughout the whole Fabrick, but what was contrived for some Use or Convenience, and with the handsomest Compactness of all the Parts.
>
> (ibid., p. 118)

Ornament is the aesthetic manifestation of nature's ground in the building's non-natural figure; its power lies in the *appearance* of natural harmony, strength, and unity it conveys. The fundamental importance of this appearance is implicit in Alberti's and Serlio's writings but only achieves full expression in the mid-sixteenth century. Palladio writes, for example: '[N]o one should leave brick walls or mantle pieces (*nappa, camino*) rough; they should be made with the greatest refinement because, apart from the misuse of materials, it would follow that *what should naturally be whole will appear broken and divided* into many parts' (1997, p. 16, my emphasis). As Alina Payne argues, ornament 'describes or re-presents what making architecture involves and in displaying a logical support system in action, literally aestheticizes strength' (1999, p. 183). Decoration is linked to decorum: properly proportioned, it produces the effect of strength (and of unity, beauty, naturalness) through the eye of the viewer. Ornament is thus essential to all buildings (Alberti, 1965, pp. 202–3), but it can never be recognized as essential; such recognition would spoil the effect of the image, make ornament visible as the source of the 'beauty' that, according to Alberti, is supposed to be innate to a structure's 'naked' body (p. 203). Ornament in proportion acts as a screen that erases the import of ornament, framing in its place a view of the power and privilege of the building's maker and its owner (ibid., pp. 186–8). Ornament out of proportion muddles the view, makes the corporeality of decoration visible. The result, Alberti argues, is 'monstrous' (ibid., p. 202).

Although Vitruvius, Alberti, and their followers overtly gender neither ornament nor the naked body it clothes (the columnar orders, for example, are based on both male and female bodies), the value systems shaping their terminology tell an implicitly sexed and gendered story. The naked body of the building is male; the decoration that pretends to reflect its beauty and power, while actually producing both, is female. For Serlio, good architecture is 'solid', 'simple', 'plain', 'smooth, gradual', 'soft, delicate' in texture; bad architecture is 'weak', 'slender', 'delicate' (as opposed to 'solid') in structure, 'mannered', 'rough' in texture, 'confused', and 'disordered' (qtd in Payne, 1999, p. 136). For Inigo Jones, whose work in England was influenced directly by Palladio, the classical exterior had to be 'masculine and unaffected' while interiors could reflect the passion and whimsy of 'nature hirself' (Anderson, 1997, p. 50). The influence of Jones's classical practice on the architectural thinking of the Stuart court meant that '[t]he connection between architecture and ideals of masculine nobility was explicit [in England] by the early seventeenth century' (ibid., pp. 50–1). 'Feminine' attributes, granted no implicit value of their own, become 'the material sign of an immaterial [male] presence' within these systems of classification, while '[t]he task of architectural theory becomes that of controlling ornament' (Wigley, 1992, p. 357), managing the fraught relationship between 'beauty' and 'ornament', the body *of* the building and the body *in* the building. The price of failure is the return of another kind of female body, empowered to ruin. Serlio writes:

> There will be for example, a beautiful and well-formed woman, who, in addition to her beauty, will be ornamented with rich vestments: *but more serious than lascivious*, and will have a beautiful jewel on her

forehead and at her ears beautiful and rich earrings: all of which add ornament to the beautiful and well-formed woman. *But if she were to have these jewels placed at her temples, and over her cheeks, and in other places* superfluamente, *tell me please, will she not be monstrous?*
(qtd in Payne, 1999, p. 140, my emphasis)

For early modern architecture, the fecund female body, the 'lascivious' female body covered by inappropriately placed 'jewels', is at once a building's structural support and the body that ruins the building, the ostentatious monster at its heart. The building is her body's container, but she is its Pandora: seeking attention, too fond of acting her central part in its larger story, she is poised always to escape, to bring down the house.

Beatrice Joanna, whose sexual availability Middleton and Rowley key explicitly to the decoration of built space both on and off her father's estate, can valuably be understood within the framework I trace above. All of her 'crimes' in the play, insofar as they implicate her 'lascivious' body, her claims to ownership over her own desire, are first and foremost crimes of architecture. Similarly, the 'punishments' that befall her – rape at De Flores's hands in 3.3; rape and murder at Alsemero's behest in the final scene – deliberately engage space in specific ways: they are designed to force her body back into a pose of architectural regulation and consolidation, to bend and break that body so that its bones may be used, once more, to shore up Vermandero's broken walls. Donald Hedrick and Bryan Reynolds have argued that *The Changeling's* 'place-in-drift' takes audiences steadily away from fixed spatial co-ordinates and toward an anxiety about placelessness consistent with changes in early modern conceptions of space and place (2006, p. 112); while their argument provides a welcome focus on the complexities of space in the play, the notion of a 'drift' away from spatial specificity does not fully account for the sexed and gendered dimensions of Renaissance architectural practice to which the play owes an important debt. Beatrice Joanna's experiences of sexual and physical violence must be made invisible precisely because her corporeal relationship to Alsemero's temple and Vermandero's fortress has become, by 3.3, *too* visible. Her rape and murder are punishments for wayward building.

The second half of *The Changeling* (assuming, as I do above, an interval set at the end of Act Three) opens with a dumbshow: audiences return from intermission to see Beatrice Joanna married to Alsemero. The action then shifts to Beatrice Joanna, alone before Alsemero's closet. She tells us that De Flores has 'undone' her 'endlessly' and that she is 'fearfully distressed' by this undoing but also by what it bodes for her own wedding night, when Alsemero expects to see and feel her virginity as his own (4.1.1–10). What follows is one of the most famous, and most commented upon, scenes in the play: Beatrice Joanna finds the key in the closet door, looks within, and discovers an alchemy manual as well as glasses marked 'C' and 'M', the contents of which supposedly prove whether a woman is pregnant or a virgin respectively. Beatrice Joanna reads the manual, learns the gestural proofs provoked by the contents of glass M, and then feeds those contents to her maid Diaphanta to watch them work. She cons Diaphanta into taking her place in Alsemero's bed on her wedding night and then uses the knowledge Diaphanta's reaction to the liquid provides to become a quite perfect dissembler of virginity. When Alsemero (inevitably) calls upon her to drink the dram herself, she 'feign[s]' 'th'effects' 'handsomely' (4.2.137–8).

Many critics have noted that Beatrice Joanna proves herself an actor in 4.1 (Burks, 1995, p. 779 and 2003, p.173; Jardine, 1996, p. 124; L. Hopkins, 2002, p. 18; Amster, 2003, p. 229; Liebler, 2003, p. 375; most famously, Garber, 1996, p. 360). Perhaps more important than *what* she does in this scene, though, is *where* she does it – and what she does *to* that place with her playing. A man's closet or private study was 'the true center' of the early modern home (Wigley, 1992, p. 348); as a trusted guest in Vermandero's house, Alsemero would have been entitled to sleeping quarters with such an inner chamber adjacent to it (ibid., pp. 347–8; Friedman, 1989, pp. 146–7). Men's closets stood as sentinels of order; they operated as 'intellectual space[s]' seemingly 'beyond sexuality' (Wigley, 1992, p. 347) that fuelled the invention of the private (male) self (Orlin, 1994, pp. 186–8; Bassnett, 2004, p. 402). But they were also essential to the shaping of female space and female identity in the house, marking 'the internal limit' to a woman's household authority (Wigley, 1992, p. 348). For Beatrice Joanna to arrive at Alsemero's closet door, she must already have been inside, or very near to, Alsemero's private chamber; this is not a space strictly barred to her as his wife, but it is also not a space universally open without express invitation. Beatrice Joanna's turn as actress, therefore, relies on a prior spatial transgression, her (temporary) reconfiguration of the household's gendered map.

Once inside the 'intellectual space' of the closet, Beatrice Joanna behaves exactly as Renaissance builders feared a lascivious ornament might do: she raises the spectre of excess, of her female body's needs, anxieties, and desires, by putting things out of order. But just as ornament, despite Alberti's occasional protests to the contrary, is less tacked-on stuff than absolutely integral to the image of the (pure, whole, ordered) body the classical façade represents (Alberti, 1965, p. 113; see also Payne, 1999, p. 75), Beatrice Joanna does not infect or invade the closet with the foreign substance of her now-sullied sexuality. Rather, her work exposes characteristics that are already innate to the closet, pulling back its curtain to reveal the corporeal contradictions that underlie the closet's subjectivating illusions. Although Alsemero's closet represents one of the most enduring constructions of private space in the Western imaginary – the idea of interiority itself – the meanings ascribed to the closet in early modern culture were actually quite varied. Lisa Hopkins points out that closets were as likely to be 'demarcated for the exclusive use of women' as for that of men, and were, 'moreover, associated with the domestic skill of food preparation' (1997, p. 149). When Beatrice Joanna substitutes the liquid in glass C with milk, she reminds us that Alsemero's closet is really a kind of kitchen (ibid., pp. 149–50; see also Boehrer, 1997), a zone of women's work. Indeed, the closet was used 'as a conventional metaphor for anatomical and spiritual interiors' and was regularly applied, in religious texts, 'to the feminine reproductive system' (Boehrer). 'To closet' also suggested the abjecting of what was impure in the body (in the sense of 'closet[ing] away' bodily residue) (Wigley, 1992, p. 344). The early modern closet was a body but also a kitchen and a toilet; it was messy interiority but also pure façade; it was a place of physical as well as intellectual labour; it was multiply gendered, heterogeneous in its contents and its functions, and yet it restricted female access on the grounds of a naturalized sexual hierarchy of space and its uses. It pretended to operate beyond sexuality, and yet it relied on an *a priori* claim to the sexual (glasses marked C and M; bodily residue). It stood as a social and psychical representation of natural order, and yet was also, at bottom, a place of performance – the performance of household labour, bodily labour; the performative production of the self.

When Beatrice Joanna and Diaphanta act their alchemical comedy in front of Alsemero's closet door, they do not turn the closet *into* a stage, proving Beatrice Joanna merely a cunning performer; they reveal the closet *already to be* a stage – a heterotopia built on bodies and their props and made for self-invention, in which Beatrice Joanna is but one performer among many. Beatrice Joanna is an actress, but she is also an architect; in fact, she is an architect precisely because she is an actress. In rehearsal with Diaphanta before the closet (and in her subsequent performance of virginity for Alsemero in 4.2), Beatrice Joanna presses her performing body into a space designed to represent Vitruvian propriety and asexuality, the male self born of man alone. As she breaks into the closet, she discovers both the forgotten female bodies (and bodily functions) and the abject theatricality on which its smoothed-out façade depends. Act Four, scene one marks a crucial turning point in *The Changeling* not because Beatrice Joanna performs, but because *the closet* does: the women's play reveals this space – and by association all the other non-spaces that work to entrap Beatrice Joanna in *The Changeling*, and for which the closet stands as sentinel – to be no more and no less than theatres, the sites of and tools for the performance of a very specific kind of self, and a very ugly kind of self-destruction.

This reading of the closet may strike some as utopic, and I agree that it advances an interpretation that seems at best temporary. By 5.3 Alsemero is again in control of his private space and Beatrice Joanna becomes its 'prisoner' (5.3.87). Nevertheless, in this climactic scene the closet yields itself up, once more, to the vicissitudes of the stage. De Flores and Beatrice Joanna disappear within to 'rehearse again / [Their] scene of lust' (ll. 115–16); the rest of the *dramatis personae* gather to hear Alsemero expose his wife's transgressions while De Flores completes the closet's promise of bodily management and purification. But they also hear something else: Beatrice Joanna's cryptic cries from behind the closet curtain or door ('Oh, oh, oh!' [l. 139]). Marjorie Garber argues that these sounds raise the disruptive spectacle of woman's disconcerting ability to 'fake it' under any circumstances, and thus of the impossibility either of knowing whether this is sex or violence, or of parsing what Beatrice Joanna 'feigns' from what she truly 'feels' (1996, p. 364). Embedded in Garber's analysis is another possibility, however: that Beatrice Joanna's 'horrid sounds' (5.3.142) are the sounds of acting, full stop – the sounds of a voice from backstage that cries out to fill in the plot. The closet, in this crucial climactic moment, is every inch both a fictional space and a performative one, a space that signs its doubled status, that reveals the role performance space, the performance *of* and *by* space, plays repeatedly both in violating Beatrice Joanna and in covering up that violation.

Renovating Beatrice Joanna's closet: the Cheek by Jowl *Changeling*

The interval and the prospect from the garden mark the spatial co-ordinates of violence's disavowal in *The Changeling*; in 5.3 that disavowal returns, uncannily, in Alsemero's closet, a very specific place we all see with our eyes and hear with our ears, but that none of us may see through or into. Space and place are malleable things the political promise of which can be made differently in performance (Tompkins, 2006, pp. 164–5; D.J. Hopkins, 2008, p. 32; Miller, 2006); Alsemero's potion cupboard, as both early modern closet and thoroughly modern theatre, offers contemporary producers an ideal opportunity to stage the violent politics of early modern spacing, but also to manipulate the often covert politics of contemporary theatrical watching. Not every *Changeling* will seize this opportunity, of course, and I'd be lying if I said I've seen many productions that even tried. It is far too easy to stage a titillating 5.3, with an inconspicuous closet and a straightforward emphasis on the resolutions taking place in the foreground. Under the right circumstances, Beatrice Joanna's eerie cries of 'Oh' can lose their ambiguity, even sound as though Alsemero orchestrates them (5.3.140, 142–3). When such a Beatrice Joanna emerges from the closet to abject herself before her male relations (ll. 149–61), text and performance collude to direct an audience's gaze away from the closet's opacity, away from its prisoner as a victim of violence, and toward the relief an end finally brings.

Olivia Williams, however, was no such Beatrice Joanna. And Cheek by Jowl's was no such *Changeling*.

Nick Ormerod's industrial exit door contained a slim window; his closet was at all times semi-transparent. During the 5.3 climax, as Williams screamed and cried on the door's far side, Keen's De Flores pushed her body repeatedly up against its window, but the brief glimpses of frantic hair, clothes, and blood that afforded was not enough to generate a clear reading of the scene for spectators in the auditorium. Instead, the moment instantly recalled, for me, the queer visual blank that was the prospect from the garden. Hiddleston's Alsemero draped himself across the closet door, reading the scene for us as sex not violence, adultery not rape; meanwhile, the prospect from the window alternately confirmed and unsettled his claims. When Keen finally hurled Williams from the closet, she paused briefly, and languidly, in the doorway, seeming in her pose to mimic Hiddleston's embodied reading of moments earlier. Was she confirming his expectations, or mocking them? She then fell to the ground at centre stage, her belly covered in blood (Figure 7). She gave her lines of self-abjection through total exhaustion, legs like dead weights sprawled out in front of her; the heft of Beatrice Joanna's shame and desecration seemed to be too much for this performer's body. In the one moment when Beatrice Joanna must appear utterly herself, utterly 'real' in order to compel her

Figure 7 Olivia Williams as Beatrice Joanna and David Collings as Vermandero in Cheek by Jowl's 2006 production of *The Changeling*. Photo by Keith Pattison, with his kind permission.

audience's belief in the play's ending, Williams showed us the physical labour that goes into creating the cruel fiction that is Beatrice Joanna at her most frighteningly self-effacing. She gave us the closet's troubling promise: Beatrice Joanna not as intractable villain, but as the ugly product of an intractable stage.

Yet Williams did not end there. She rose with the other cast members on Alsemero's penultimate speech, moving to stand downstage centre beside Hiddleston and facing the audience. Hiddleston proclaimed 'Here's beauty changed / To ugly whoredom' (5.3.197–8), thrusted a pointed finger at Williams, and charged us to confirm his diagnosis. Williams then turned toward Hiddleston's face, gave him a dismissive, slightly bemused look, turned to give the same look to the audience, and walked calmly away.[6] Watching Williams's Beatrice Joanna refuse Alsemero's final judgment and rebuke the play's demand to end, gutted, on her knees, I could not help but think of Janet McTeer's response to Bosola's 'mercy' in Phyllida Lloyd's production of Webster's play. Both actresses rejected not only the terms of the scripts they had been handed, but also the specific *places* into which those scripts would fix them. Williams's final gesture of defiance, made from within

Donnellan's Brechtian frame, left her audiences with a Beatrice Joanna planted firmly in self-conscious performance space: looking back at the structures that enclose, define, and shape her as an invisible victim of invisible violence, but also looking out at her witnesses, charging us to do the same.

Williams's last moments on stage embodied perfectly the larger ethos of Cheek by Jowl's production, an ethos best recognized in Nick Ormerod's design as it folded the play's performance spaces 'inside out' (Clapp, 2006). The central pillar and door were the only fixed elements in the space; to one side, a desk, chair, CCTV camera, sink, and small fridge suggested backstage space, rehearsal space, the space where 'prospects' are formed and illusions are perfected.[7] During the show's London run, which will be the main focus of my analysis below, the company stretched this interpretation to extremes. Cheek by Jowl turned the Barbican's vast, industrial backstage area into its playing space; spectators, meanwhile, crammed into semicircular, makeshift seating pushed several feet up onto the stage. The vast and infinitely more comfortable Barbican auditorium sat empty behind them, eliciting yowls of protest from a number of critics. These protests – along with some far less angry, and far more nuanced, reactions – attest to the equally uncomfortable questions this production raised about the role space (and comfort, luxury, institutional hierarchy) plays in the production of audience desire, power, and control over the meaning of the spectacle that is Beatrice Joanna. The exit sign above the closet door suggested not a secret space to be breached or discovered on its other side, but rather that audiences, along with the actors, were already *inside*, the keepers of the closet and its secrets, responsible for its power, responsible to its victims, acutely aware of its theatrical dimensions. Actors and spectators alike found themselves marooned in unfamiliar territory, displaced from the traditional hierarchies of 'monumental' theatre architecture. Performers assumed the pose of spectators on orange plastic chairs when not acting on a stage coded as a workspace. Audiences sat in an uncanny double of the auditorium behind them but also on the Barbican's stage, finding themselves materially implicated in the performance and (by many accounts of the stiff, tiny seats) literally pained by the labour of watching, both throwing into relief the relations of commerce and capital on which the hierarchies of the theatre (usually silently) depend (Ridout, 2006, p. 10 and *passim*). In fact, the only thing that appeared to be missing from this *Changeling* was the stage itself, the world of seamless, pleasurable illusions that usually makes its heroine such a compelling vixen.

This production was not about *The Changeling per se*; it was about how *The Changeling* gets made in the modern theatre. It was about the stakes of creating and consuming Beatrice Joanna, generating a psychological thriller about sexual violence as sexual desire, psychic trauma as personal shame, and compelling its belief. In her 2006 *Sunday Telegraph* review, Susan Irvine called the production's engagement with the play's pseudo-Freudian heritage 'part homage, part parody'; as Donnellan noted in a 2006 interview with Domenic Cavendish, his goal was to create 'an immediate and sensual emotional connection' in audiences while simultaneously provoking questions about 'what that [emotional connection] means'. The extensive metatheatrical frame he and Ormerod placed on the story accomplished two things, each of which I will explore below: first, it drove a rereading of Beatrice Joanna through the lens of plastic production, the meticulous building of a character in performance and the specific coercions that generate particular responses to that character in an audience; second, as it inverted not just the world of performance (rehearsal hall for stage) but also the realm of our watching (stage for auditorium), it encouraged a rereading of the play through the spaces of its contemporary production and reception. While only the Barbican show placed its spectators directly on the stage, every performance on the tour featured the purpose-built rehearsal hall set, allowing all audiences to see the play at least to some degree through the processes of its theatrical making.

Madness and acting

Olivia Williams was the heart of Donnellan and Ormerod's challenge to *Changeling* convention. She complicated any audience desire for, or attempts at, a typical Freudian reading of her character's relationship with De Flores by investing in a very different psychological trajectory for Beatrice Joanna. Williams pitched her performance as a blend of manic anxiety and barely contained terror from its first moments, and held it there until her final scene; she thus not only refused Beatrice Joanna's expected character journey from innocence to sexual awakening, but also refused the misogynist modern commonplace that Beatrice Joanna achieves emotional complexity only through her attachment to De Flores. Moreover, her Beatrice Joanna was neither a naïve sex kitten nor a simple victim of De Flores's unwanted sexual advances; she was a longtime victim, and survivor, of sexual abuse at her father's hands. Williams told this back-story subtly but with unmistakable force early in the performance, allowing it to saturate Beatrice Joanna's future interactions with De Flores and every other male character on the stage.

At the top of 2.1 Beatrice Joanna employs Jasperino as a go-between, handing him a letter instructing Alsemero to visit her; between his exit and De Flores's entrance she soliloquizes her good 'judgement' in her choice of lover and proclaims that she 'see[s] the way to merit, clearly see[s] it' (2.1.13–14). Typically this soliloquy offers the play's first opportunity to indict Beatrice Joanna as at best ignorant and at worst manipulating and cavalier, but in this production it marked the spot where Beatrice Joanna's uncomfortable bodily and emotional history collided with the text. The actors playing Vermandero, Alonzo, Alsemero, and several others remained on stage, scattered around the space, frozen in position; as she spoke Williams travelled from body to body, laying her words on heads, arms, shoulders. On 'he's so forward too' she arrived behind Vermandero (David Collings), who stood in turn behind Alonzo (Laurence Spellman). In Middleton and Rowley's text the lines 'he's so forward too, / So urgent that way, scarce allows me breath / To speak to my new comforts' (ll. 24–6) refer either to Vermandero's or to Alonzo's eagerness to conclude the arranged match between the latter and Beatrice Joanna – and quite possibly to both. Williams, however, transformed the lines with body language that suggested this forwardness – and her expected obeisance – referred to her father's repeated molesting of her. Pausing briefly, Williams swallowed a sob on 'forward', inclining her head toward Collings's shoulder. On 'So urgent that way' she pushed her hips toward him with a sneer, then moved away, crying, her face (as ever in this production) in her hands. This scene was one of several that evolved considerably over the course of the production's tour, increasingly emphasizing Beatrice Joanna's violent and emotionally troubled past; even from the earliest performances in Nancy, France, though, the history of rape was palpable. In Nancy, Williams cupped her right hand near Vermandero's bottom on 'so forward'; on 'So urgent' she reached toward his head as though to grip it in a vise, then backed away. Collings's Vermandero supported Williams's reading of Beatrice Joanna's history by treating his daughter as a pimp would a whore: as he promised Alsemero a chance to see his castle and sample its entertainments, he circled Williams and gripped her around the waist suggestively.

Even as I advance the above argument and attribute its choices directly to Williams, I am aware that I may be treading on unstable ground. In an interview with Domenic Cavendish in March 2006 Williams expressly rejected the argument that Beatrice Joanna is raped by De Flores, and suggested instead that her attraction to him is horrifying and confusing to her because he is disabled. While I do not wish to imply that Williams's fundamental understanding of her character shifted over the course of the tour, I do want to suggest that more may have been going on in her complex construction of this character than she was willing or able to account for publicly, either at this early point or later, in May 2006, when she wrote an article on the tour for *The Independent*. By appearing to set Beatrice Joanna's choices and reactions throughout the play within the context of a history of sexual abuse, Williams injected a contemporary feminist sensibility into her character's early modern context; more importantly, she wrote the history of rape (and of rape's forgetting, its normalization in Vermandero's household) on Beatrice Joanna's body and on her emotional past, present, and future. Given early modern culture's tendency to understand sexual violence primarily as men's experience, and given modern *Changeling* performance's tendency to understand Beatrice Joanna's experiences with De Flores as both singular and originary (not to mention unproblematically consensual), Williams's apparent decision to tie her erratic actions throughout the play to a personal history of sexual trauma carried important political weight as well as cultural risk. This decision had another welcome consequence, too: it complicated Beatrice Joanna's status as an actor.

From 2.2, when Beatrice Joanna first pretends affection for De Flores in order to secure his service, to 4.2, when she dissembles her virginity for Alsemero and Jasperino by aping Diaphanta, to 5.1 when she must perform surprise and heartbreak at Diaphanta's death, Williams was comically awkward; her 'acting' was laden with emotional instability and was even, at times, obviously reluctant. Her virginity test in 4.2 consistently elicited laughter from audiences because her performance seemed so plainly contrived, but its contrivance barely masked her turmoil. At the Barbican Williams sat in her chair, numbed; she fought back tears amid the laughter, and finally broke down, face once more in her hands, sobbing with relief at the sadness that marks the final phase of the test. In Spain, late in the tour, Williams resisted any touch from Hiddleston during this scene; her gapes and sneezes were forlorn, her laughter cut short by fatigue, her 'joy of heart' (4.2.144) contradicted by a hand that fluttered to her temple. Williams's 'bad' acting worked at the limits of the play's antitheatrical obsession with Beatrice Joanna (L. Hopkins, 2002, pp. 27–8): the lines between her own will to perform, her need to play to others' expectations, and her psychic confusion was never clear. This woman acted her obedience – to her husband, to her father, to De Flores – because she had no other choice, because obedience had always meant acting, and thus acting, even with tremendous

Figure 8 Olivia Williams as Beatrice Joanna in Cheek by Jowl's 2006 production of *The Changeling*. Photo by Keith Pattison, with his kind permission.

difficulty and amidst constant psychic eruptions, was simply a matter of survival (Figure 8).

Any psychological reading of Beatrice Joanna ultimately risks an all-too-familiar reception practice; Williams's reading, however progressive it might have been, was no exception. Donnellan and Ormerod mitigated this risk by locating Williams's work very clearly within their larger metatheatrical conception of the play, and specifically within the heterotopic blend of madness and acting articulated by the production's playing space. That space was not simply a rehearsal hall; the exit sign, the surveillance camera, bare-bones desk, and institutional-grade plastic chairs layered connotations of 'mental hospital' seamlessly on top of the more pervasive denotations of 'backstage space' (Edwardes, 2006; Peter, 2006). When the scene shifted from Vermandero's castle to Albius's madhouse, lights came up to a full, intense white, recalling an examination room but also casting a glare on the space's every nook and cranny, giving it more than ever the feel of a working theatre.

Freud himself appeared in the subplot, but as a character (Jim Hooper's Albius) quite clearly playing Freud, complete with 'egghead beard and little round glasses' (Irvine, 2006); Williams, meanwhile, played Beatrice Joanna playing a woman in a madhouse during the first subplot scene (1.2), and again immediately before the final act. The dance in which Antonio and Franciscus meet at the end of Act Four quickly morphed into a musical number by the entire cast, which entered through the closet door; the number had a Hora-like, celebratory quality about it, both manic and contrived, jumbled yet too precise, the spectacle not of madness but of a group of people pressed to perform an image of charming mania for Lollio's expectant spectators. David Benedict (2006) complained that the actors in the mad scenes 'collectively look[ed] like a bunch of actors doing mad acting', but surely this was the whole point. Beatrice Joanna, in this scene and all the others, was at once a Bedlam resident and a performer of bedlam; as a psychological case study she was wholly vexing because she always appeared both as a sufferer of clearly readable sexual trauma *and* as a member of a theatrical team, both producer and product of theatrical desire. She performed because she was ill, she performed to ward off illness, but she also performed because madness, in this play, is born of consumer demand and (as Benedict helpfully demonstrates) spectatorial urging. Her prison was a madhouse in which interpretations are fixed and people are trapped by others' readings of their inner lives; it was also a rehearsal room in which interpretational shifts become possible, and cultural shifts (hopefully) begin to take place.

The tension between 'feigns' and 'feels', madness and acting, in Williams's performance was particularly fraught in her several encounters with Keen's De Flores. In Middleton and Rowley's text these scenes are characterized by a substantial number of asides; the resulting intimacy between actor and spectator creates the impression that we have access to the characters' innermost thoughts, not to mention their intentions. In this production the asides carried an added burden: they signalled De Flores's phantasmatic projections of his own desires onto Beatrice Joanna's body, and thus generated parallel uncertainty about the authenticity of what we were seeing of, and on, that body. Lighting designer Judith Greenwood and sound designer Gregory Clarke marked De Flores's entrances to Beatrice Joanna with sudden, dramatic shifts: a low clang signalled the beginning of one of his asides, while a spotlight marked him off from the rest of the stage. Asides straddle the space between stage and audience, pulling spectators ever more tightly into the orbit of performance; this production heightened and politicized that effect by placing De Flores clearly on the limen between theatrical worlds. He was foremost a member of the audience, the world of watching, but one who nevertheless commanded a remarkable power to affect materially the trajectory of the performance.

The sound and lighting shifts at De Flores's entrances momentarily froze the stage, giving him control over it for the duration of his aside. At the Barbican, he began his entrance to Beatrice Joanna in 2.2 at the top of the audience risers, stage right; his spot caught him as well as several spectators while he moved slowly down to performance level to stand behind Beatrice Joanna's chair. As he spoke his expectation that any woman who 'fl[ies]' from her husband inevitably 'spreads and mounts then like arithmetic – / One, ten, a hundred, a thousand, ten thousand' (2.2.61–3), Williams arched backward, running her hands across her neck, chest, breast and legs, materializing his words on her body much as Frankford's words tore across Saskia Reeves's broken frame in Mitchell's *Woman Killed*. Was Beatrice Joanna dreaming Alsemero's body on hers? Was she foreshadowing her response to De Flores's advances? Or was she simply acting out De Flores's lines, a puppet of his directorial pull? Was this genuine sexual hunger, the phantasmatic body of early modern misogyny, or the body born of modern reception desire? There was no way to know for sure: each possibility seemed equally plausible.

This array of options meant that audience members could, if they so wished, easily choose to read Williams's gestures as burgeoning sexual expression, and the body of reviews from the London run suggests plenty of professional spectators were unwilling to let go of their preferred reading of the character. As the play progressed, however, those same audience members may have found it increasingly hard to ignore the tension building between the stuff of De Flores's fantasies, painted onto Beatrice Joanna's performance body, and her resistance to them, manifested through a body traumatized by a history of sexual violation. At the climax of 3.3, as De Flores pressed his suit for sex, Keen became increasingly agitated; he began throwing chairs while Williams sought cover behind one and then another, him chasing her the length and breadth of the Barbican stage and wing areas. The effect was startling: even in the hugeness of this space Beatrice Joanna could find no place to hide. If the stage was the hated closet rendered inside out, its interior vastness was for her no more capacious than the head of a pin, and its theatrical promise served only to heighten De Flores's directorial power over the scene. Finally Keen grabbed Williams from behind and began running his hands across her body in sinister mimicry of her earlier self-touch, bringing the violence of his earlier fantasy projection literally to life. Williams's Beatrice Joanna, tired and broken, opened her hands as Keen roughly kneaded her breasts; fighting sobs, she stumbled on her confused words of surrender, then brought her hands once more to her face. She looked like she was about to vomit.

'Shroud your blushes in my bosom' (3.3.166), De Flores tells his prey, and I'll read your silence for consent until you learn to love me (ll. 167–70). Keen hoisted Williams front-ways onto his pelvis on 'shroud your blushes'; suddenly, she had her back to the audience. The rest of his words filled in the resulting visual gap with his dream of her love – just as Jasperino's words would do later, filling out the prospect from the garden with his expectation of Beatrice Joanna's betrayal. But this moment's ambiguity would not last: as Keen turned to carry Williams toward the desk, her face came back into focus. Her eyes looked like she was struggling to stay conscious and aware; she breathed hard into Keen's shirt through gritted teeth, like a terrified child. Keen laid her on the desk and began to rip off her panties; Williams arched her back, just as she had done during De Flores's aside in 2.2, but this time the echo was distorted, his fantasy body openly competing with the violence impacting her real one. As Williams curled backward she reached for and knocked over the desk chair, still sobbing; she brought her hands to her face and forehead, grunting in pain. Then, as Keen ripped her legs open, the lights suddenly came down with a thunderous boom, signalling the end of De Flores's control over the stage. He and Williams rose and walked calmly off through a bright shaft of light, past the audience risers at stage left. The scene had been his rape fantasy, and it exposed the vicious ends of his obsession with watching and reading Beatrice Joanna, the extraordinary power his privileged spectatorship held over her body in pain. And more: as Keen and Williams disappeared through the light that marked the unexpected return of their material stage, they left spectators to grapple with an interval that placed the violent potential of reception desire squarely in its spotlight.[8]

From this fraught moment forward, Williams's 'fallen' Beatrice Joanna both met and defied the conventional expectations De Flores's fantasies produce for offstage spectators. The result was an awesome demand on audiences to make meaning out of *our* reactions to her reactions in the space between 'feigns' and 'feels' – a task that became increasingly challenging as the performance wore on and the ambiguity of Williams's actions increased. Then, during 5.1, Williams finally took from Keen the reins of the stage.

Act Five, scene one is pivotal for proponents of the 'romantic' reading of *The Changeling*. In the heat of her panic over the wedding night's dissembling, with Diaphanta playing virgin in her husband's bed, Beatrice Joanna praises De Flores's 'care' of her and calls him 'a man worth loving' (l. 71, 76), apparently confirming that rape has yielded love and devotion after all. And yet Beatrice Joanna also tells De Flores she is 'forced' to love him, and that she is moved by 'fear' to conform to his demands (l. 48, 58). The underlying ambiguity is political; it speaks exactly to what cannot be said, or seen, of the factors mitigating Beatrice Joanna's attachment to De Flores. Williams played this ambiguity with brutal force. '[D]o what thou wilt now' (l. 36) she cried at Keen, pulling him toward her against the closet door and putting his hand on her breast; she adopted a distinct roughness on 'I'm forced to love thee now, / 'Cause thou provid'st so carefully for my honour' (l. 48–9), giving the line a bitter, cruel edge. No longer was Beatrice Joanna in thrall to De Flores; she had come to realize that playing his fantasy whore was the only means through her present hell and she used this new knowledge to appeal directly to her audience, inviting us to see what she could not clearly show us, to read the terror of emergency hidden between her lines. Keen set fire to Diaphanta's chamber; Williams pulled the fire alarm next to the closet door, rousing the house at the top of her lungs. Keen reappeared; he swung Williams, seemingly to both their joys, in his arms. Then quickly the sound and light changes used to mark De Flores's asides echoed through the space; this time, however, the stage belonged not to Keen but to Williams. 'How rare is that man's speed! / How heartily he serves *me*!' (l. 69–70, my emphasis) she sneered over the din of the alarm, arms in the air, standing at centre stage powerful and defiant. 'Look upon his care' (l. 71) she shouted directly at the audience, rubbing once more her breasts and belly at the thought of De Flores, but now with ironic affectation. Keen, meanwhile, appeared in a shaft of bright light stage right, frozen in mid-run, arrested by the sudden force of Beatrice Joanna as she owned at last her own performance.

The politics of theatrical discomfort

This Beatrice Joanna was ever both a cultural and a theatrical product. Williams bought into the trend to psychologize the character, but not in the way seasoned audiences may have expected; she both capitulated to and resisted the Freudian projections that screened her body, offering spectators choice and challenge rather than the ready-made image of our favourite early/modern bad girl. And in this challenge and choice – in the prevailing uncertainty over the difference between Beatrice Joanna's desires, De Flores's desires, and our own; in the production's sometimes uncomfortable requirement that we decide who this woman is, and then watch as her actions unravel our expectations – Cheek by Jowl interpellated its audiences as key players within its working performance space. But the company's spatial interventions in this *Changeling* also went much further. As Donnellan and Ormerod physically manipulated the Barbican stage, they also asked audience members to consider how they come to the theatre, understand their place within a specific theatre's range of artistic practices, and take advantage, by association, of that theatre's cultural capital in order to make meaning of performance for their own benefit.

Nicholas Ridout writes:

> [...] in the modern theatre, something of our relationship to labour and to leisure is felt every time the theatre undoes itself around the encounter between worker and consumer. When we experience something of theatre's 'ontological queasiness', pry too deep into the backstage darkness, or witness the whole apparatus turning itself inside out, we are experiencing something of this political relationship. While it may always have been thus, it is experienced all the more intensely now, under conditions of capitalist modernity.
>
> (2006, p. 34)

The 'ontological queasiness' that shapes Ridout's argument was the Cheek by Jowl *Changeling*'s defining affect. The production's power to trouble its audiences lay in its tendency, as I have been arguing, to thwart expectations: over who Beatrice Joanna is, and how she is *for us* tonight; over what a 'proper' auditorium looks and feels like; over where the backstage stuff should properly be stashed. But in that thwarting, the production also materialized, uneasily, the uneven social and economic relations that motor the modern, classical-realist theatre. The problem with that theatre, Ridout argues, is that it 'does its work while you watch. In the theatre you always know you are there, at the scene of the action, at the site of production. Seeing yourself there, and others there, and facing up to the nature of your relationships with these others, is what disquiets the mind and degrades the art' (ibid., p. 29). Normally we don't see ourselves there, of course; we see the stage, the actors see each other or the footlights, and the hierarchies of power and privilege that define each group's relation to the other disappear in the spaces between. Normally.

The Barbican Centre is one of London's premier arts and culture venues, and according to its online promotional material is 'the largest multi-arts centre in Europe'. Located in the middle of the City of London, just steps from the headquarters of major banks and international financial institutions, the Barbican occupies a position of elite privilege within London's psychic geography yet also demonstrates a populist, even socialist ethic. Sustained through an amalgam of public and private money, the Barbican embodies the best of the social welfare state in the twenty-first-century West: it is accessible to anyone in the London area by excellent transit links, offers a range of events in the arts, music, film, and theatre for diverse audiences at various price points, and represents the full force of (especially local) government support for the arts. Its Associate Companies include both traditional cultural organizations like the London Symphony Orchestra and the BBC Symphony Orchestra, and high-profile avant-garde companies like Cheek by Jowl and the Michael Clark Dance Company. Until 2002, the Barbican was also the London home of the Royal Shakespeare Company. Despite this apparently seamless blend of tradition and innovation, however, the Barbican ultimately favours partnerships with artists and groups that have already achieved excellence in the eyes of the nation's arts and media establishments. It views itself as a purveyor of 'quality' British art, both at home and abroad. As Artistic Director Graham Sheffield (2006) writes, the companies with which the Barbican associates 'all boast a fine international reputation and a proven ability to bring the best of ideas from Britain to the world, both in traditional virtues, as well as in contemporary practice and presentation'.

Cheek by Jowl's play with the Barbican's physical shape challenged these enabling terms and conditions. Donnellan and Ormerod turned the site's 1200-seat main theatre from a traditional proscenium, one of the Centre's anchor venues, into a site-specific performance space and their *Changeling* into a commentary on the social architectonics of spectatorship at a large, elite, modern theatre within the elite spaces of a 'global' city (McKinnie, 2009). In the process the company asked of itself and of its spectators the question Lois Keidan argues is central to all live art: 'what is the place of this kind of work in this kind of place?' (2006, p. 11). In order to inhabit the stage of the Barbican Theatre, Cheek by Jowl had first to confront the legacy of the RSC, its predecessor in the space and one of the most respected theatre companies of its kind in the world. The RSC had built a set of expectations at the Barbican over the years, in both the main house and in the smaller, studio-sized Pit theatre: large budgets, excellent actors, and reliable stagings of classic plays. Cheek by Jowl's rejection of the plush, wide Barbican house in favour of a makeshift seating area, coupled with its embrace of the unvarnished stage and backstage areas, immediately signalled a break with these expectations, a refusal to let pass unremarked the power of both stage *and* auditorium space to authorize and naturalize a specific, 'correct' version of the classics. For a number of critics this break proved a failure on two grounds. They argued, first, that the shift into bleachers and onto the stage sacrificed audience comfort and, second, that it prevented rather than encouraged emotional connection between actors and audience. Nicholas de Jongh (2006) complained that the 'Fan-shaped banks of seats [gave] a sense of limbo yet outlaw[ed] scenic intimacy' (see also Shuttleworth, 2006; Brown, 2006; and Berkowitz, n.d.). Michael Coveney (2006) charged that the company's reduction of the auditorium was 'an admission of defeat', as though *their* kind of work in this kind of place would never draw enough spectators to fill the storied main house, and they well knew it. Charles Spencer (2006) went so far as to suggest that the seating reconfiguration was an affront to the audience, turning 'one of the most comfortable and well-designed theatres in London into one of the worst' (see also Letts, 2006).

These parallel complaints – about discomfort on one hand, and failed intimacy on the other – go a long way toward revealing the conditions under which theatrical spectacle in this 'monumental' space typically empowers elite audience members like professional reviewers and subscription patrons. To be comfortably ensconced is more than to be able to enjoy the show: comfort marks the spectator as clearly different from, and in an important sense superior to, the actors. Their labouring bodies on stage signal theatre as *work* for them and *play* for us, the work of reviewing notwithstanding; the pleasure we take in this distinction makes the whole evening worth the price of admission, and lends the critic his or her discriminating voice. Comfortable seats are the *sine qua non* of the monumental theatre's hierarchy of space: they make us 'gentles' and the actors our humble servants – but they also don't force us to talk or think about this defining distinction. 'The actor is always at risk of "yielding [...] privacy", in the act of self-revelation that is psychological acting', writes Ridout, 'while the spectator is permitted to enjoy the feeling of intimacy that comes from witnessing acts of self-revelation in others, without disclosing anything of herself, safe in the darkness of her seat' (2006, p. 77). Comfortable seats in darkened auditoriums disappear beneath our bodies, allow our bodies themselves to disappear, to be absorbed into the stage action without a backward glance; this is why, for a critic like de Jongh (2006), the tighter Cheek by Jowl configuration prevented rather than encouraged 'scenic intimacy'. The vast Barbican auditorium, his comments imply, will shrink on its own if you'll just let it be; each spectator will soon enough find him or herself alone in a tight little room with De Flores and Beatrice Joanna. If, on the other hand, you make an issue of the auditorium, no such spatial collapse will be possible; audiences will be too aware of the architectural, and perhaps thus also the political and economic, dynamics mediating their enjoyment of the scene.

To enter Cheek by Jowl's *Changeling* at the Barbican, audiences had to follow a path 'through a side door, into [the] wings[,] then on through the pitch black to the bowels of the stage' where a 'scaffold' had been erected, 'staring into a vast brick vault with weirdly angled walls' (Bassett, 2006). Kate Bassett called this topsy-turvy world 'a disorienting found space within a familiar theatre'. While a number of critics praised the company's set-up for the ways in which it mirrored the claustrophobic worlds of the play, several followed Bassett's lead in understanding the audience, rather than the actors or the characters, as the main subjects of Donnellan and Ormerod's spatial disorientation (see also Shenton, 2006; and Allfree, 2006). The Cheek by Jowl *Changeling* interrupted the conventional dynamics of production and consumption at work at the Barbican, making spectators' tacit dependence upon a certain kind of viewing space in such a theatre one of the tangible subjects of its staging. Asking audiences to think every step of the way to their seats, to experience the *work* of spectatorship in a physical, embodied way throughout the performance, the company drew sustained attention to the typically invisible links between stage and auditorium as parallel spaces of making, parallel worlds in which theatrical pleasure arises from the careful privileging of a particular view, framed at a particular angle, and supported by parallel acts of willed erasure.

* * *

The Cheek by Jowl *Changeling* activated the heterotopic potential embedded within Vermandero's fortress walls and laid the groundwork for an activist practice of watching Beatrice Joanna. Donnellan and Ormerod layered the secretive spaces of Middleton and Rowley's text onto the vast expanse of the Barbican stage and auditorium, revealing how Beatrice Joanna, fetish object, is made in the collusion of two offstage worlds: the safe space of our viewing, and the terrorizing fictional interiors that disappear from our sight only to enable her bodily and emotional trauma. The Barbican's main house became kin to the prospect, the interval and the closet in this production; like those cloistered non-spaces, the vast, comfortable array of seats tucked out of sight behind Ormerod's makeshift risers telescoped the seamless collusion of architecture, narrative, and imaginative process through which violence becomes sex, trauma becomes love, and coercion becomes betrayal at the scene of Beatrice Joanna's undoing. Of course, Beatrice Joanna was still called a whore; of course, Cheek by Jowl were still chastised for daring to meddle so grossly with the Barbican. It would be disingenuous of me to claim that this *Changeling* irrevocably changed the trajectory of the play in performance. But it did set a fresh, and necessary, precedent for the future, interrupting the pleasure of the view just long enough to throw the meanings of this story, in this space and time, into question.

Afterword

In the climactic scene of Richard Eyre's 2004 film, *Stage Beauty*, Maria (Claire Danes) plays Desdemona to Ned Kynaston's blacked-up Othello (Billy Crudup). Kynaston was formerly the most famous Desdemona on the Restoration stage, and Maria was his dresser; now she is the new 'it' girl, and Ned is out of work and out of favour. Perhaps for love, perhaps for money, Ned agrees to coach Maria's Desdemona and to take the part of Othello in a performance of the play that represents continued professional survival for them both; at the end of this performance, however, the tensions bisecting Ned's motives boil over. In a scene the DVD titles 'Method Acting', Danes and Crudup deliver pitch-perfect psychological realism that veers off course as Ned's rage toward Maria overtakes his actor's instincts. He throws her around, roughly forces her onto the bed, and finally smothers her with shocking brutality. On the scene's flip side, between her urge to act well and her genuine terror Maria's performance begins to tip over the footlights and into the Real. 'Help! [...] He's killing me!', she cries desperately between lines of text, between pre-rehearsed actions; audience members grow increasingly edgy, stirring in their seats, but nobody makes a move to interrupt. Finally, Crudup drops the pillow; he stumbles away into a corner while Danes remains on her back, apparently suffocated. The camera pans to his face – confusion and terror – and then to the faces of distinct spectators: Tom Wilkinson's Betterton, Hugh Bonneville's Samuel Pepys, Zoe Tapper's Nell Gwynn, and Rupert Everett's King Charles II. Everyone is frightened; no one knows what to do.

After what seems like minutes Maria sputters back to life, and Nell and Charles herald the audience's relief by clapping their bravos. Nell has Maria's number: this has all been an act, and the audience has figured out the trick (and approves). Or has it? Moments later Maria, exhilarated, tells Ned: 'You nearly killed me!' He replies, 'I did kill you. You just didn't die.' This exchange provokes, at last, the conscious recognition of Ned and Maria's entirely conventional, heterosexual romance, preordained from the film's first moments (despite its promising investment throughout in the politics of Ned's queer persona); it suggests too a new, and preferred, method of 'being' on the early modern stage. Stanislavski has miraculously supplanted late seventeenth-century declamation, and no actresses were hurt in the making of this thespian revolution. We can all go home now, safe and well in the knowledge that our intervention was not required, and that Desdemona lives to die another day.

And yet our intervention *was* called for. The eruption of modern method technique is only half of this climactic scene's fantasy; the other half envisions a body of spectators suspended on the cusp of Roger Simon's 'encounter with difficult knowledge', that which might bring them more than they know what to do with (2005, p. 10). In the endless moment of uncertainty between Maria's smothering and her sputtering return, questions cram the air. What if she is dead? What do we do now? Do we rush the stage? Should we help her? How? The camera scans the auditorium; its focus is as much on the spectators as the actors. What responsibility will they take for this woman, this life? Can the conventions of performance – them up there, us down here, a chasm in between – really be so hard and fast as to prevent, to rebuke, necessary humanitarian intervention? The applause that greets Maria's reawakening is of course only partly for her and Ned; it is also, in the main, for the audience itself, a grateful substitute for the much riskier action no spectator could bring him or herself to take. A standing ovation masks a collective refusal to intervene, a collective refusal to recognize, to encounter, the very real violence rained down on the woman on stage beneath the thin veneer of theatrical business-as-usual.

I end here because *Stage Beauty*'s climax so elegantly braids this book's two principal strands. First, it demonstrates with particularly modern force (the force of 'method acting') how performance functions in early modern England to render women's experiences of violation invisible in plain sight, actionable (perhaps) in extreme circumstances yet also strangely unrecognizable in those same circumstances as violence to be acted upon *for a woman's sake*. (At the same time, in the audience's awkward suspension between the threat of a 'real' death and the return of performance, the scene reveals the extent to which the theatre can still function today to make a fetish of violence's invisibility, to draw perverse pleasure from the spectre of a woman's genuine pain.) Second, the scene encodes, even if it does not enact, the responsibilities to which contemporary performances of women's violation can, and should, call audiences. Our stages remain filled with scenes just like this one, incredible scenes of violence against women that are very hard to watch but surprisingly easy to rationalize away.

I have been preoccupied in these pages not just with a performative history of violence's erasure at one particularly potent moment in English theatrical production, but also, and especially, with the uptake of that performative history today. The productions I have explored invoke the possibility of making a fetish of their violated female bodies, only to settle on a call to audience 'response-ability' and 'addressability' (Oliver, 2001, pp. 5–7) for the suffering woman at the centre of the scene, a call for true recognition of the damaged other in her profound alterity. Lavinia loses her hands and tongue and cannot complete rape's metatheatrical return; Sonia Ritter and Laura Fraser, their work separated by more than a decade, each find a way to plumb the depths of that loss, to stage its inaccessibility in the face of the stage's overwhelming urge to render it meaningful beyond Lavinia's body. Anne Frankford takes the curse of her husband's 'kindness' in at her mouth and starves herself to death; Saskia Reeves and Michael Maloney will not let us forget the physical brutality this seemingly benevolent word enables, their warped, rasping, gulping voices demanding our visceral encounter with its unspoken underbelly. The Duchess of Malfi's performance of despair is mocked by Bosola's martyrology and subsequent revenge; Lucy Peacock, Joyce Campion, and Janet McTeer return her body to the stage with the full force of its missing witness, staging the Duchess's loss, in Act Five, as a loss we in the stalls enact with our averted eyes, our refusal to play the part of the politically empathetic spectator the Duchess demands. Finally, Beatrice Joanna experiences a sexual violation that, even before it happens, cannot be named; Olivia Williams charges us to see marked on her body a history of sexual violation that offers an alternative psychological trajectory for Beatrice Joanna, even as Nick Ormerod and Declan Donnellan force audiences into a radical new physical relationship with the character and her stage.

Yet the stories I tell of these productions are often tinged with resistance from professional theatre critics, leaving me uncertain of the productions' ultimate public impacts. Reviewers ignored Ritter, as her own male relations did (to ironically comic effect) in the latter half of Warner's *Titus Andronicus*; they similarly resisted Mitchell's *Woman Killed* and Hinton's and Lloyd's *Duchess*es, often reserving some of their loudest disdain for the female leads. More reviewers than we might at first expect appreciated Cheek by Jowl's *Changeling*, but few fully understood the ramifications of the company's interventions – especially its extraordinary manipulation of the play's viewing spaces – and chose instead to enjoy Beatrice Joanna in a manner all too familiar to London audiences. I have often argued in these pages that professional reviewers cannot be taken as avatars of the broad theatregoing public, but the truth is their voices carry far more popular influence than mine. In the wake of this rather uncomfortable admission, I cannot help but wonder: can, will, a feminist performance of violence against women on the contemporary early modern stage ever be fully recognized as such? Can we move beyond our still-pervasive urge to sanctify the Bard as politically flawless even as we misread the feminist in everyday representation as flawed, as hysterical, as an affront to spectatorship itself? Can we come to terms with the profound political potential of what artists like Warner, Taymor, Mitchell, Hinton, Lloyd, Donnellan and Ormerod are trying to accomplish?

And so this Afterword is not a rehearsal of the historical arguments I have developed over the course of this book; it is not a rewriting of the production analyses I have rendered. Instead, it is a call to arms for all who have in the past called themselves feminist spectators, and who are willing now and in the future to call themselves feminist spectators. It is a call to see 'beyond recognition', to accept the challenge to our preferred ways and means of knowing embedded in those performance moments in which we do not know what to do, where to look, how to feel. Our natural urge as spectators in such moments is to shut down, to resist, to brand the performance lacking; but what if it is *our* lack that the performance would show us, our susceptibility to the seductive lure of the cover-up that we ought to be seeing? Let us go now to watch early modern performance prepared to see the gaps between text and speech, word and action; let us go prepared to encounter the differences, as well as the similarities, between the historical and the now, the ghosts in theatre's machine. Let us go prepared to interrogate the feminist potential of this work, its political engagement with human history's extraordinary investment in violence against women, and to record those interrogations in writings for a broad public. I hope that I have been able to show something of what an ethical, feminist performance of violence against women in contemporary early modern theatre can look like; let us go now, together, to make its ethical reception a reality.

Notes

1 Encounters with the Missing: From the Invisible Act to In/visible Acts

1 The focus on rape's historical effacement is in no way limited to scholarship of the Renaissance. See Gravdal (1991) and Roberts (1998) for the medieval perspective; see the work collected in Higgins and Silver's *Rape and Representation* (1991), especially Joplin (1991); and see Smart (1989), Estrich (1995), and de Lauretis (1987) for contemporary views of rape's continued elision in twentieth-century discourse.
2 Karen Bamford's *Sexual Violence on the Jacobean Stage* (2000) includes an extensive, thoroughly researched survey of early modern rape law in its introduction; as a result, I do not rehearse all the law's details here.
3 Bamford (2000) focuses in particular on the rapes of virgins in the drama of the period, and on the ways in which rape operates, paradoxically, as a proof of chastity otherwise not visible to the naked eye. Deborah Burks handles the flip side of the problem: the cultural management of the rapes of women whose sexual loyalty is suspect. Burks's analysis of Middleton and Rowley's *The Changeling* (1995) demonstrates the internal paradox on which the early modern effacement of rape hinges: the crime is at once not real violence against a woman, and yet a very real act of violence that a woman suspected of sexual duplicity may commit against the men to whom she is bound.
4 Jill Dolan, Peggy Phelan, Alice Rayner, Joseph Roach, David Román, Rebecca Schneider, and Diana Taylor (whom I discuss below) have done some of the most exciting recent work in the sub-discipline of performance-as-witness; its pioneer, however, is Herbert Blau (1982).
5 For examples of feminist film criticism heavily invested in the violence of the gaze, see E. A. Kaplan (1983), Freedman (1990), and Bronfen (1996). Teresa de Lauretis's *Technologies of Gender* (1987) was the first text in the field, as far as I know, to explicitly state 'that violence is engendered in representation' (p. 33) as it linked 'the violence of rhetoric' with the production of gender difference (p. 47).

2 Rape's Metatheatrical Return: Rehearsing Sexual Violence in *Titus Andronicus*

1 'Ravished' is the direction for Lavinia's entrance at the top of Act Two, scene three in the 1995 edition of the play (ed. Bate) I cite here; some other editions give this scene as 2.4.
2 In addition to Lucrece and Lavinia, consider Webster's Virginia (*Appius and Virginia*), Middleton's Lady (*The Second Maiden's Tragedy*), Fletcher's Lucina (*Valentinian*), and Dekker and Massinger's Dorothea (*The Virgin Martyr*). For a comprehensive analysis of the 'classical paradigm' in Jacobean

representations of rape, see Bamford, 2000, esp. pp. 61–80. By 1606 the narrative trajectory and iconography of this style of 'return' was familiar enough to warrant parodic treatment in *The Revenger's Tragedy*, in which Antonio's wife is discovered post-rape and post-suicide clutching not one prayer book but two, opened to messages proclaiming her innocence.

3   The exception to this rule was young women: the presumed sexual innocence of the virgin, ironically, gave her the freedom to speak of sexual assault in much more graphic terms (Gowing, 2003, pp. 94–5).

4   The editions of Glanvill and Bracton I use offer side-by-side Latin and English translations, allowing me to compare both the original and the translations for shifts in terminology and emphasis. In addition to being much more extensive than Glanvill's, for example, Bracton's advice to rape victims includes the phrase 'hue and cry' ('clamore et huthesio'), while the earlier tract does not.

5   The problem of rape's 'effacement' in medieval and early modern literature and culture has been an issue for feminist scholars since at least the 1970s. For early work on the topic, see Kahn (1976), Stimpson (1980), Carolyn Williams (1993), Joplin (1991), and Gravdal (1991); more recent work includes Baines (1998), Roberts (1998), Bamford (2000), and Burks (1995 and 2003). Pascale Aebischer's chapter on *Titus Andronicus* in her *Shakespeare's Violated Bodies* thoughtfully explores 'the implications for performance' and 'for literary criticism' of 'the textual gap left by Lavinia's erasure' in the play (2004, p. 25), but does not take the kind of historicist approach to rape's effacement that I aim for here.

6   This approach to *Titus* begins in the 1970s with articles by Palmer (1972), Tricomi (1974), and Hulse (1979), but gains strength from the force of deconstruction's influence on the British and American academies through the 1980s. See Fawcett (1983), Kendall (1989), and Rowe (1994).

7   In addition to Fawcett and Kendall, see Marion Wynne-Davies (1991) and Karen Robertson (2001). Emily Detmer-Goebel (2001) offers a historically nuanced take on the Lavinia-as-author debates in 'The Need For Lavinia's Voice: *Titus Andronicus* and the Telling of Rape.'

8   A number of critics have suggested that Lavinia's writing constitutes a re-creation of her violation; see Bott, 2001, pp. 202, 204; Marshall, 2002, pp. 202, 205; and D. Green, 1989, p. 325.

9   Freud revisited the castration moment several times in his writings on human sexuality; various accounts can be found in 'Three Essays on the Theory of Sexuality' (1905), 'On the Sexual Theories of Children' (1908), 'The Dissolution of the Oedipus Complex' (1924), 'Some Psychical Consequences of the Anatomical Distinction Between the Sexes' (1925), and 'Femininity' (1933).

10   My remarks here are based on my viewing of the Swan Theatre production of the play on archive video in July 2006. The performance was recorded on 27 August 1987.

11   Amanda Konradi writes: 'Since the 1970s, a set of physical and cognitive symptoms has become recognized as the impact stage of rape trauma syndrome. A woman whose sense of self is fundamentally shaken is often unable to speak in complete sentences or to convey whole thoughts. She may appear incoherent, disorganized, and dazed. If her sense of time is also

      warped by rape, planning future action can be difficult or even impossible' (2007, p. 34).
12. My thanks to my student Greta Fairhead for this observation.
13. Most mainstream and alternative media reviews of *Titus* focus on Taymor's gruesome aesthetic, either debating its strengths and weaknesses or panning it outright. While a few reviewers critique the acting, very few mention Fraser; the elision is not jarring, however, as Taymor does not concentrate on Lavinia in the same way Warner sought to do. (Stephen Holden, writing for the *New York Times*, makes an exception; he calls the moment of Lavinia's discovery '[t]he film's most searing image' (1999), suggesting a visual experience both resonant and haunting.) I do not consider mainstream reviews to be a particularly effective index of Fraser's (or the rape's) reception, nor does my discussion seek to claim Taymor's film as a feminist success. Rather, I am interested in how her techniques might be deployed in future feminist performances of sexual violence.
14. Spears'...*Baby One More Time*, with its disturbing undertones of violence-as-sexual-play, was released in early 1999; *Titus* premiered on 25 December 1999.
15. Lisa Hopkins offers an excellent summary of the use of tiger imagery throughout the play, and compares it to Taymor's augmented use in the film (2003, pp. 61–2).

3 The Punitive Scene and the Performance of Salvation: Violence, the Flesh, and the Word

1. While there is little surviving documentary evidence of the life of Phillip Stubbes, we do know Katherine's son John was baptized on 17 November 1590; the Parish register of Burton-upon-Trent also records the death of Katherine in the final weeks of 1590 (Kidnie, 2002, pp. 4–5).
2. 'In all her sickness, which was both long and grievous, she never shewed any sign of discontentment or of impatience' (Stubbes, 1992, p. 144); 'She was accustomed many times as she lay, very suddenly to fall into a sweet smiling and sometimes into a most hearty laughter, her face appearing right fair, red, amiable, and lovely' (p. 145); moments before her death she sings 'certain psalms most sweetly and with a cheerful voice' (p. 148).
3. Susan Amussen's historical data suggests that neighbours, friends, and community leaders would have routinely intervened to help battered women when authorities failed to do so; still, her evidence also points to serious problems with this safety net. See 1994, especially pp. 78–9.
4. The shift away from physical correction in the conduct literature is in part a matter of class: gentlemen don't hit their wives. See Dolan, 1996, p. 15; Detmer, 1997, p. 278; and Woodbridge, 2003, p. xiii.
5. Historians generally agree that the late sixteenth and early seventeenth centuries saw a rise in household disruptions, and in wives suing for separation from abusive husbands. For a contemporary account of a wife's court action, see Crawford and Gowing, 2000, pp. 173–5. For women's collective efforts to persuade the judiciary to recognize excessive violence, see Mendelson and Crawford, 1998, p. 216.

6   While companionism and mutual subjection are not new ideas, and in fact can be traced back to marital ideals in place before the Reformation (Davies, 1981, p. 78), what is new in the later sixteenth century is 'the emphasis, elaboration, and wide distribution' these ideas 'receive in the Puritan tracts' (Rose, 1988, p. 119; see also Davies, 1981, pp. 61, 58).
7   Whately stands explicitly opposed to Gouge on this matter; compare *Of Domesticall Duties* (1976), pp. 389–93.
8   The *OED* traces the modern definition of 'kindness' (that is, the quality of being kind) to approximately 1350; it also notes, however, that as early as 1400 kindness could mean 'a natural inclination, tendency, disposition, or aptitude'. In 1536 the term was first used to denote a 'natural right or title derived from birth or descent'.
9   *Woman Killed* was written in 1603 and first published in 1607; it remained popular and came out in a third edition in 1617, the same year Whately first published *A Bride-Bush*. While Heywood obviously could not have had *A Bride-Bush* in mind when he first wrote the play, we can safely assume that both texts take advantage of ideas about 'kindness' circulating in England in the first two decades of the seventeenth century.
10 Fasting to extremes had long been a matter of concern for Church authorities; the saintly woman who took her asceticism too far could easily be read as resisting rather than miming religious and social doctrine. Caroline Walker Bynum notes that by the later Middle Ages the traditional cyclical fasts of early Christianity had given way to a doctrine of moderation, in which leaders and thinkers as diverse as Bonaventure and Thomas Aquinas argued that spiritual rather than physical abstinence was the true path to God (1987, pp. 42–5). Fasting subsequently became a genuinely resistive practice for many spiritual women. Both Nancy Gutierrez (1994) and Frey and Lieblein (2004) follow Bynum's lead in reading Anne's fast as a form of social and religious resistance.
11 Mitchell's *Woman Killed* opened at The Other Place, Stratford-upon-Avon, in 1991 and transferred to The Pit in London in the spring of 1992. The archive video I watched in July 2006 was recorded at The Pit on 21 May 1992.
12 Uncertain about whether the darkness I was seeing on the archive video was part of the production's design or an accident of recording technology, I spoke with Sylvia Morris, Head Librarian at the Shakespeare Centre, about her experience watching the play at The Other Place. Morris characterized that experience as a sheer struggle to see through the constant dim.
13 While Freud found much to use in Ibsen and others, I do not wish to imply that naturalist realism is in any way based in Freud. Method techniques pioneered in America under Lee Strasberg tend to rely heavily on mid-century psychology, but Stanislavski was more interested in psycho-physical processes than in psychoanalysis. Moreover, while psychological realism has evolved into a performance technique problematic to feminism for a number of reasons, it is in no way antithetical to feminist theory or performance practice – as Mitchell's work ably demonstrates. See Blair (2002) and Gainor (2002) for two nuanced feminist reassessments of Stanislavski.
14 I saw *Iphigenia at Aulis* twice: once live in July 2004, and again on archive video at the RNT archives on 18 July 2006. The video production I watched was recorded on 1 July 2004. All quotations from the production are taken from the video recording, not the prompt script.

4 Witness to Despair: The Martyr of Malfi's Ghost

1. Act Five rarely offers much fodder for feminist analysis. See, for example, Rose, 1988, p. 171; Callaghan, 1989, p. 96; Jankowski, 1992, p. 181; Haber, 1997, p. 147; Daileader, 1998, p. 91. Several recent readings, however, have begun to reverse this trend. Barbara Correll argues that the Duchess's death leaves the remaining male actors 'unmoored, affectively and socially' (2007, p. 91), while Reina Green (2003, pp. 467–8) and Gina Bloom (2007, pp. 160–1) offer provocative readings of the Duchess's fifth-act return as a form of unstable female vocal agency.
2. For a similar take on Webster's relationship to his audience, see Diehl, 1997, p. 185. While Diehl's carefully historicized examination of the play intersects with my own at the point of the witness, I diverge from her reading as I consider the difficult contours of the Duchess's relationship to Protestant performance tradition (pp. 196–8).
3. In my epigraph, 'them' refers not to the martyrs but to 'the stars', which the Duchess curses on 4.1.95–6. The passage, however, invites more than one interpretation. The Duchess's last reference to the stars appears prior to Bosola's intervention on line 106; set off alone, the lines I quote make perfect sense as an independent clause.
4. Margaret Owens has shown that the early post-Reformation period featured representations of active, powerful, female Protestant martyrs, but by the late sixteenth century the tone had shifted, and women like Lady Jane Grey appeared in print as less self-determining and far more humble (2005, p. 131).
5. Cynthia Marshall calls 'the drama of martyrdom' 'a pleasure derived from the three interlocking dialectics of (de)valuing the flesh, promoting/erasing individuality, and strategically collapsing the domains of word and deed' (2002, pp. 89, 102). This is the very pleasure the Duchess takes pleasure, and pain, in performing against in Act Four, scenes one and two.
6. For a smart reading of Bosola's performance investments, see Barker, 2005.
7. Patricia Phillippy details College of Arms regulations governing what roles women could or could not play during formal funerary rites in the period, even – and especially – when those women were close to the deceased (2002, pp. 21–3). These regulations all suggest patriarchal anxiety over the power a keening woman's body might exert in male social space.
8. In their historical survey of *Malfi* productions, McLuskie and Uglow note that, as early as 1611, the King's Men may have had access to a device that would have allowed the Duchess to emerge as a fully embodied ghost in a burst of light in 5.3 (1989, p. 11). Such a ghostly spectacle works very differently from Hinton's echo chain. It reincarnates the Duchess just as the performance of salvation attempts to imagine her: as whole, physically untouched, purified, saved.
9. Like Mitchell's *Woman Killed*, this production was notoriously dark. A large number of reviewers disdained this choice, suggesting it hampered audience views of characters' facial expressions. See, for example, Coulborn, 2006; Portman, 2006; Scowcroft, 2006; Smith, n.d.; Hoile, n.d.; and H. Simpson, 2006. By contrast, Robyn Godfrey (2006) argued that the low lighting actually amplified the skull effect throughout.

10 On the material pressures that shape Stratford performance, see Knowles, *Shakespeare and Canada* (2004b), chs 1 and 2, and Knowles, *Reading the Material Theatre* (2004a), pp. 105–28.
11 I watched Lloyd's production on archive video over two days in mid-July 2006. I saw Hinton's production live twice that same summer, and also watched it on archive video in March 2008 and June 2009.
12 The Duchess was not the only onstage observer during Act Five: every member of the cast whose character was killed rose to deliver, as McTeer had done, a detached final line, and then joined her on the risers. These other observers were peripheral to the scene of witness Lloyd constructed, however; they sat largely motionless and in shadow while McTeer's acts of watching were brightly lit and carefully animated.
13 Hinton, ironically, received virtually the same critiques. Like Hinton, both Lloyd and McTeer suffered from reviewers' conventional urge to weep for the Duchess, and thus to understand any other approach to the play or the role as a betrayal of its sympathetic impulse. In addition to Billington, 2003, see Coveney, 2003; Edwardes, 2003; Foss, 2003; Macaulay, 2003; Nightingale, 2003; Peter, 2003; and Wolf, 2003. By contrast, several reviewers praised McTeer for providing appropriate emotion despite Lloyd's best efforts to squelch a naturalist reading of the Duchess (Brown, 2003; de Jongh, 2003; Gore-Langton, 2003). Charles Spencer (2003) sums up the reviewers' collective anxiety when he claims that McTeer 'never came close to moving [him], appearing eerily detached from a role that *requires* an actress to spill her emotional guts all over the stage' (my emphasis).

5 The Architecture of the Act: Renovating Beatrice Joanna's Closet

1 I viewed three different archive DVD recordings of this production (from early, middle, and late points in the tour) at Cheek by Jowl's Barbican Centre offices in late November 2007. I owe sincerest thanks to Jacqui Honess-Martin and the rest of the office staff for their generosity both with archive materials and with personal impressions. The productions I viewed were recorded at: Nancy, France in mid-April 2006 (no date on the DVD recording); the Barbican Centre, London on 5 or 6 June 2006; and Almagro, Spain on 1 July 2006. My reading of the production will derive primarily from the Barbican performance, unless otherwise noted.
2 Theatre critics are not the only writers who routinely manufacture Beatrice Joanna's consent, and encourage audiences to do the same: academic writing on the play is often similarly guilty. Editors' introductions to published editions of the play are, especially for prospective student audiences, more likely to have a serious impact on spectatorial expectation than newspaper reviews, and both recent New Mermaids student editions of the play (Daalder, 1990; and Neill, 2006) advocate explicitly for Beatrice Joanna's sexual complicity. For several other recent examples of critical writing that adopts a version of the 'romantic' reading of the play, see Stockholder, 1996; Daalder and Telford, 1999; Neill, 2005; and Sugimura, 2006.

3 Baines, citing *Lawes Resolutions* 396, notes that another man's concubine or 'whore' could have been recognized as raped if she actively resisted her rapist; in other words, her status as 'whore' did not extend beyond the property line of the man who claimed her (1998, pp. 77–8).
4 I am deliberately generalizing, of course; I realize that not all productions of *The Changeling* set their intervals at the end of 3.3, even though it marks the climax of the first half of the action and a natural stopping point midway through the play. My analysis of 'the interval' in this chapter assumes that the majority of productions do, and will, set their intermissions as De Flores pulls or carries Beatrice Joanna off stage; it also, however, aims to make a case for why future productions of this play interested in politicizing Beatrice Joanna's experience *should* consider setting their intervals here.
5 A number of feminist architecture theorists have explored the missing female body in the history of classical and modern architectural practice. See Ingraham, 1998, especially 'The Outline of the Dead Body', as well as Bergren, 1992 and 1996; Bloomer, 1993; and Grosz, 1995. On the performance dynamics of 'woman as housed', see Colomina, 1992.
6 Not all audiences will have seen this provocative ending: archive recordings suggest that, early in the tour, Williams responded to Hiddleston's accusations of whoredom by turning quickly away, as in shame. On the archive DVD of the Barbican Centre performance, Williams gives Hiddleston the look, but does not offer it to the audience; the performance recorded at the Almagro Festival a month later, however, includes both looks.
7 Jacqui Honess-Martin, Cheek by Jowl's associate producer at the time of this production, told me that the door and pillar into which it was set were inspired by a door the actors used for entrances and exits in the company's actual rehearsal space. While Honess-Martin notes that Ormerod did not intend a particular reading of the space with this choice, the 'backstage' connotations were quite clear, especially when paired with the company's manipulation of the Barbican stage in London.
8 While by no means all reviewers assessed this scene as rape, David Benedict (2006), Timothy Ramsden (2006), and even the typically conservative Charles Spencer (2006) all commented on the 'explicit' violence of the moment (Ramsden).

Bibliography

Aebischer, Pascale. (2001). Review of *Titus*, by Julie Taymor. 1 February. *Scope: An Online Journal of Film and TV Studies*. Institute of Film Studies, University of Nottingham. 28 July 2008. Available at: <http://www.scope.nottingham.ac.uk/filmreview.php?issuefeb2001&id818&sectionfilm_rev)>.
——. (2004). *Shakespeare's Violated Bodies: Stage and Screen Performance*. Cambridge: Cambridge University Press.
Agustí, Clara Escoda. (2006). 'Julie Taymor's *Titus* (1999): Framing Violence and Activating Responsibility.' *Atlantis: revista de la Asociación Española de Estudios Anglo-Norteamericanos*, 28.1: 57–70.
al-Solaylee, Kamal. (2006). 'What's missing? Passion!' *The Globe and Mail*, 6 June, p. R3.
Alberti, Leon Battista. (1965). *Ten Books on Architecture*. Trans. Cosimo Bartoli and James Leoni. Ed. Joseph Rykwert. London: Tiranti.
——. (1988). *On the Art of Building in Ten Books*. Trans. Joseph Rykwert, Neil Leach, and Robert Tavernor. Cambridge, MA: MIT Press.
Allfree, Claire. (2006). 'Bordering Madness and Sanity.' Review of *The Changeling*, by Thomas Middleton and William Rowley. *Theatre Record*, 7–20 May, p. 578. (Rpt. from *Metro (London)*, 17 May 2006.)
Amster, Mara. (2003). 'Frances Howard and Middleton and Rowley's *The Changeling*: Trials, Tests, and the Legibility of the Virgin Body.' *The Single Woman in Medieval and Early Modern England: Her Life and Representation*. Ed. Laurel Amtower and Dorothea Kehler. Tempe, AZ: Center for Medieval and Renaissance Studies, pp. 211–32.
Amussen, Susan Dwyer. (1994). '"Being Stirred to Much Unquietness": Violence and Domestic Violence in Early Modern England.' *Journal of Women's History*, 6.2: 70–89.
——. (1995). 'Punishment, Discipline, and Power: The Social Meanings of Violence in Early Modern England.' *Journal of British Studies*, 34.1: 1–34.
Anderson, Christy. (1997). 'Masculine and Unaffected: Inigo Jones and the Classical Ideal.' *Art Journal*, 56.2: 48–54.
Baines, Barbara J. (1998). 'Effacing Rape in Early Modern Representation.' *ELH*, 65.1: 69–98.
Bal, Mieke. (2001). 'Calling to Witness: Lucretia.' *Looking In: The Art of Viewing*. Amsterdam: G+B Arts International, pp. 93–116.
Bamford, Karen. (2000). *Sexual Violence on the Jacobean Stage*. New York: St. Martin's Press.
Barbican Centre. (n.d.). 'About Barbican: Overview.' www.barbican.org.uk. 14 March 2008. Available at: <http://www.barbican.org.uk/about-barbican>.
Barker, Roberta. (2005). '"Another Voyage": Death as Social Performance in the Major Tragedies of John Webster.' *Early Theatre*, 8.2: 35–56.
——. (2007). *Early Modern Tragedy, Gender, and Performance, 1984–2000: The Destined Livery*. Basingstoke: Palgrave Macmillan.

Barker, Roberta, and David Nicol. (2004). 'Does Beatrice Joanna Have a Subtext? *The Changeling* on the London Stage.' *Early Modern Literary Studies* (online), 10.1. Available at: <http://extra.shu.ac.uk/emls/10-1/barknico.htm>.
Bashar, Nazife. (1983). 'Rape in England Between 1550 and 1700.' *The Sexual Dynamics of History: Men's Power, Women's Resistance.* London: Pluto Press, pp. 28–42.
Bassett, Kate. (2006). 'Lost in the labyrinth.' Review of *The Changeling*, by Thomas Middleton and William Rowley. *Theatre Record*, 7–20 May, p. 577. (Rpt. from *The Independent on Sunday*, 21 May 2006.)
Bassnett, Madeline. (2004). ' "A Frightful Pleasure, That is All": Wonder, Monstrosity, and The Changeling.' *Dalhousie Review*, 84.3: 387–406.
Bate, Jonathan. (1995). Introduction. *Titus Andronicus*, By William Shakespeare. Ed. Jonathan Bate. London: Routledge, pp. 1–121.
Bawcutt, N. W. (1998). Introduction. *The Changeling*. By Thomas Middleton and William Rowley. Ed. N. W. Bawcutt. Revels Student Editions. Manchester: Manchester University Press, pp. 1–29.
Behling, Laura L. (1996). ' "S/He Scandles Our Proceedings": The Anxiety of Alternative Sexualities in *The White Devil* and *The Duchess of Malfi*.' *English Language Notes*, 33.4: 24–43.
Belsey, Catherine. (1999). *Shakespeare and the Loss of Eden: The Construction of Family Values in Early Modern Culture*. Basingstoke: Macmillan.
Benedict, David. (2006). Review of *The Changeling*, by Thomas Middleton and William Rowley. *Variety*, 29 May – 4 June, p. 46+.
Bennett, Lyn. (2000). 'The Homosocial Economies of *A Woman Killed With Kindness*.' *Renaissance and Reformation*, 24.2: 35–61.
Bennett, Susan. (1996). *Performing Nostalgia: Shifting Shakespeare and the Contemporary Past*. London: Routledge.
Bergren, Ann. (1992). 'Architecture Gender Philosophy.' *Strategies in Architectural Thinking*. Ed. John Whiteman, Jeffrey Kipnis, and Richard Burdett. Cambridge, MA: MIT Press, pp. 8–46.
——. (1996). 'Female Fetish Urban Form.' *The Sex of Architecture*. Ed. Diana Agrest et al. New York: Abrams, pp. 77–96.
Berkowitz, Gerald. (n.d.). Review of *The Changeling*, by Thomas Middleton and William Rowley. *TheatreguideLondon* (online). 19 November 2007. Available at: <http://www.theatreguidelondon.co.uk/reviews/changeling06.htm>.
Bevington, David M. (1968). *Tudor Drama and Politics: A Critical Approach to Topical Meaning*. Cambridge, MA: Harvard University Press.
Billington, Michael. (1987). Review of *Titus Andronicus*, by William Shakespeare. *Theatre Record*, 7–20 May, pp. 636–7. (Rpt from *The Guardian*, 14 May 1987.)
——. (1988). Review of *Titus Andronicus*, by William Shakespeare. *Theatre Record*, 1–14 July, p. 906. (Rpt from *The Guardian*, 6 July 1988.)
——. (2003). 'Malfi, madness and melancholy.' *Theatre Record*, 1–28 January, p. 72. (Rpt. from *The Guardian*, 29 January 2003.)
Blair, Rhonda. (2002). 'Reconsidering Stanislavsky: Feeling, Feminism, and the Actor.' *Theatre Topics*, 12.2: 177–90.
Blau, Herbert. (1982). *Take Up the Bodies: Theater at the Vanishing Point*. Urbana: University of Illinois Press.
Bloom, Gina. (2007). *Voice in Motion: Staging Gender, Shaping Sound in Early Modern England*. Philadelphia: University of Pennsylvania Press.

Bloomer, Jennifer. (1993). *Architecture and the Text: The (S)crypts of Joyce and Piranesi*. New Haven, CT, and London: Yale University Press.
Boehrer, Bruce. (1997). 'Alsemero's Closet: Privacy and Interiority in "The Changeling."' *Journal of English and Germanic Philology*, 96.3: 349–68.
Bott, Robin L. (2001). '"O, Keep Me From Their Worse Than Killing Lust": Ideologies of Rape and Mutilation in Chaucer's *Physician's Tale* and Shakespeare's *Titus Andronicus*.' *Representing Rape in Medieval and Early Modern Literature*. Ed. Elizabeth Robertson and Christine M. Rose. New York: Palgrave Macmillan, pp. 189–211.
Bracton, Henry de. (1968). *On the Laws and Customs of England*. Vol. 2. Trans. Samuel E. Thorne. Cambridge, MA: Harvard University Press/Belknap.
Brison, Susan J. (2002). *Aftermath: Violence and the Remaking of a Self*. Princeton, NJ: Princeton University Press.
Bromley, Laura G. (1986). 'Domestic Conduct in *A Woman Killed With Kindness*.' *SEL*, 26.2: 259–76.
Bronfen, Elisabeth. (1996). 'Killing Gazes, Killing in the Gaze: On Michael Powell's *Peeping Tom*.' *Gaze and Voice as Love Objects*. Ed. Renata Saleci and Slavoj Žižek. Durham, NC, and London: Duke University Press, pp. 59–89.
Brown, Georgina. (2003). Review of *The Duchess of Malfi*, by John Webster. *Theatre Record*, 1–28 January, pp. 74–5. (Rpt. from *The Mail on Sunday*, 2 February 2003.)
——. (2006). 'Sex, sin and lust – but no passion.' Review of *The Changeling*, by Thomas Middleton and William Rowley. *Theatre Record*, 7–20 May, pp. 577–8. (Rpt. from *The Mail on Sunday*, 21 May 2006.)
Burks, Deborah G. (1995). '"I'll Want My Will Else": *The Changeling* and Women's Complicity With Their Rapists.' *ELH*, 62.4: 759–90.
——. (2003). *Horrid Spectacle: Violation in the Theater of Early Modern England*. Pittsburgh: Duquesne University Press.
Butler, Judith. (2004). *Precarious Life: The Powers of Mourning and Violence*. New York: Verso.
Bynum, Caroline Walker. (1987). *Holy Feast and Holy Fast: The Religious Significance of Food to Medieval Women*. Berkeley and Los Angeles: University of California Press.
Caldwell, Ellen. (2003). 'Invasive Procedures in Webster's *The Duchess of Malfi*.' *Women, Violence, and English Renaissance Literature: Essays Honoring Paul Jorgensen*. Ed. Linda Woodbridge and Sharon Beehler. Tempe, AZ: Arizona Center for Medieval and Renaissance Studies, pp. 149–86.
Callaghan, Dympna. (1989). *Woman and Gender in Renaissance Tragedy: A Study of* King Lear, Othello, The Duchess of Malfi *and* The White Devil. Atlantic Highlands, NJ: Humanities Press International.
Carlson, Marvin. (1989). *Places of Performance: The Semiotics of Theatre Architecture*. Ithaca, NY: Cornell University Press.
——. (2001). *The Haunted Stage: The Theatre as Memory Machine*. Ann Arbor: University of Michigan Press.
Casey, Edward S. (1997). *The Fate of Place: A Philosophical History*. Berkeley: University of California Press.
Catty, Jocelyn. (1999). *Writing Rape, Writing Women in Early Modern England: Unbridled Speech*. Basingstoke: Macmillan.

Chakravorty, Swapan. (1996). *Society and Politics in the Plays of Thomas Middleton*. Oxford: Clarendon.
Chaudhuri, Una. (1995). *Staging Place: The Geography of Modern Drama*. Ann Arbor: University of Michigan Press.
Chaytor, Miranda. (1995). 'Husband(ry): Narratives of Rape in the Seventeenth Century.' *Gender and History*, 7.3: 378–407.
Clapp, Susannah. (2006). 'Drowning his sorrows in style.' Review of *The Changeling*, by Thomas Middleton and William Rowley. *Theatre Record*, 7–20 May, p. 577. (Rpt. from *The Observer*, 21 May 2006.)
Cleaver, Robert. (1598). *A Godlie Forme of Householde Government: For the Ordering of Private Families, according to the direction of Gods word*. London: Printed by Felix Kingston for Thomas Man.
Coats, Catharine Randall. (1992). *(Em)bodying the Word: Textual Resurrections in the Martyrological Narratives of Foxe, Crespin, de Bèze and d'Aubigné*. New York: Peter Lang.
Coddon, Karin S. (2000). 'The *Duchess of Malfi*: Tyranny and Spectacle in Jacobean Drama.' 1993. *The Duchess of Malfi: Contemporary Critical Essays*. Ed. Dympna Callaghan. New Casebooks. Basingstoke: Palgrave Macmillan, pp. 25–45.
Colomina, Beatriz. (1992). 'Intimacy and Spectacle: The Interiors of Adolph Loos.' *Strategies in Architectural Thinking*. Ed. John Whiteman, Jeffrey Kipnis, and Richard Burdett. Cambridge, MA: MIT Press, pp. 68–88.
Comensoli, Viviana. (1996). *'Household Business': Domestic Plays of Early Modern England*. Toronto: University of Toronto Press.
Correll, Barbara. (2007). 'Malvolio at Malfi: Managing Desire in Shakespeare and Webster.' *Shakespeare Quarterly*, 58.1: 65–92.
Coulborn, John. (2006). 'Even Hell Needs Heaven.' *Toronto Sun*, 5 June. Peter Hinton/*Duchess of Malfi* clippings file. Stratford Shakespeare Festival Archives, Stratford, ON.
Coveney, Michael. (2003). Review of *The Duchess of Malfi*, by John Webster. *Theatre Record*, 1–28 January, p. 73. (Rpt. from *The Daily Mail*, 29 January 2003.)
——. (2006). Review of *The Changeling*, by Thomas Middleton and William Rowley. *Theatre Record*, 7–20 May, p. 575. (Rpt. from *The Independent*, 16 May 2006.)
Crawford, Patricia and Laura Gowing, eds. (2000). *Women's Worlds in Seventeenth-Century England*. London: Routledge.
Creaser, Sharon. (2005). 'Public and Private Performance of Guilt in Thomas Heywood's *A Woman Killed With Kindness*.' *Dalhousie Review*, 85.2: 285–94.
Cushman, Robert. (2006). 'Royal Lady Pulls Audience Over to the Dark Side.' *National Post* (Toronto), 9 June, p. PM 13.
Daalder, Joost. (1990). Introduction. *The Changeling*. By Thomas Middleton and William Rowley. Ed. Joost Daalder. New Mermaids. London: A & C Black, pp. xi–xlvii.
Daalder, Joost, and Antony Telford. (1999). '"There's Scarce a Thing but is Both Loved and Loathed": *The Changeling* 1.1.91–129.' *English Studies*, 80.6: 499–508.
Daileader, Celia. (1998). *Eroticism on the Renaissance Stage: Transcendence, Desire, and the Limits of the Visible*. Cambridge: Cambridge University Press.
Davies, Kathleen M. (1981). 'Continuity and Change in Literary Advice on Marriage.' *Marriage and Society: Studies in the Social History of Marriage*. Ed. R. B. Outhwaite. London: Europa, pp. 58–80.

Davis, Lloyd, ed. (1998). *Sexuality and Gender in the English Renaissance: An Annotated Edition of Contemporary Documents.* New York and London: Garland.

Davis, Tracy C. (1 June 2008). 'When is Theatre History?' Plenary lecture. CATR Annual Conference, Vancouver, B.C.

De Jongh, Nicholas. (2003). Review of *The Duchess of Malfi*, by John Webster. *Theatre Record*, 1–28 January, p. 72. (Rpt. from *The Evening Standard*, 29 January 2003.)

———. (2006). Review of *The Changeling*, by Thomas Middleton and William Rowley. *Theatre Record*, 7–20 May, p. 575. (Rpt. from *The Evening Standard*, 16 May 2006.)

De Lauretis, Teresa. (1987). *Technologies of Gender: Essays on Theory, Film, and Fiction.* Bloomington: Indiana University Press.

Desmet, Christy. (2000). '"Neither Maid, Widow nor Wife": Rhetoric of the Woman Controversy in *The Duchess of Malfi*.' 1991. *The Duchess of Malfi: Contemporary Critical Essays.* Ed. Dympna Callaghan. New Casebooks. Basingstoke: Palgrave Macmillan, pp. 46–60.

Dessen, Alan C. (1989). *Titus Andronicus.* Manchester and New York: Manchester University Press.

Detmer, Emily. (1997). 'Civilizing Subordination: Domestic Violence and *The Taming of the Shrew*.' *Shakespeare Quarterly*, 48.3: 273–94.

Detmer-Goebel, Emily. (2001). 'The Need For Lavinia's Voice: *Titus Andronicus* and the Telling of Rape.' *Shakespeare Studies*, 29: 75–92.

Diamond, Elin. (1989). 'Mimesis, Mimicry, and the "True-Real."' *Modern Drama*, 32.1: 58–72.

———. (1997a). 'Brechtian Theory/Feminist Theory: Toward a Gestic Feminist Criticism.' *Unmaking Mimesis: Essays on Feminism and Theater.* London: Routledge, pp. 43–55.

———. (1997b). *Unmaking Mimesis: Essays on Feminism and Theater.* London: Routledge.

———. (2003). 'Modern Drama/Modernity's Drama.' 2000. *Modern Drama: Defining the Field.* Ed. Ric Knowles, Joanne Tompkins, and W. B. Worthen. Toronto: University of Toronto Press, pp. 3–14.

Diehl, Huston. (1997). *Staging Reform, Reforming the Stage: Protestantism and Popular Theater in Early Modern England.* Ithaca, NY: Cornell University Press.

Dolan, Frances E. (1994a). *Dangerous Familiars: Representations of Domestic Crime in England, 1550–1700.* Ithaca, NY, and London: Cornell University Press.

———. (1994b). '"Gentlemen, I have one more thing to say": Women on Scaffolds in England, 1563–1680.' *Modern Philology*, 92.2: 157–78.

———, ed. (1996). *The Taming of the Shrew: Texts and Contexts.* Boston, MA: Bedford.

Dolan, Jill. (2005). *Utopia in Performance: Finding Hope at the Theater.* Ann Arbor: University of Michigan Press.

Donnellan, Declan. (9 March 2006). Interview with Domenic Cavendish (online). 27 November 2007. Podcast available at: <http://www.cheekbyjowl.com/productions/thechangeling/podcast.html>.

Edwardes, Jane. (2003). Review of *The Duchess of Malfi*, by John Webster. *Theatre Record*, 1–28 January, p. 71. (Rpt. from *Time Out*, 5 February 2003.)

Edwards, Christopher. (1988). Review of *Titus Andronicus*, by William Shakespeare. *Theatre Record*, 1–14 July, p. 902. (Rpt from *The Spectator*, 29 July 1988.)
———. (2006). Review of *The Changeling*, by Thomas Middleton and William Rowley. *Theatre Record*, 7–20 May, p. 578. (Rpt. from *Time Out*, 24 May 2006.)
Enders, Jody. (2004). 'The Spectacle of the Scaffolding: Rape and the Violent Foundations of Medieval Theatre Studies.' *Theatre Journal*, 56.2: 163–81.
Estrich, Susan. (1995). 'Is It Rape?' *Rape and Society: Readings on the Problem of Sexual Assault*. Ed. Patricia Searles and Ronald J. Berger. Boulder, CO: Westview, pp. 183–93.
Euripides. *Iphigenia at Aulis*. Dir. Katie Mitchell. Lyttelton Theatre, London: Royal National Theatre. First performance: June 2004. Archive Video Recording. National Theatre Archives, London.
Fawcett, Mary Laughlin. (1983). 'Arms/Words/Tears: Language and the Body in *Titus Andronicus*.' *ELH*, 50.2: 261–77.
Flather, Amanda. (2007). *Gender and Space in Early Modern England*. Woodbridge, Suffolk: Royal Historical Society.
Fletcher, Anthony. (1995). *Gender, Sex and Subordination in England 1500–1800*. New Haven, CT, and London: Yale University Press.
Forte, Jeanie. (1992). 'Focus on the Body: Pain, Praxis, and Pleasure in Feminist Performance.' *Critical Theory and Performance*. Ed. Janelle G. Reinelt and Joseph R. Roach. Ann Arbor: University of Michigan Press, pp. 248–62.
Foss, Roger. (2003). *What's On*, 5 February, p. 53. Phyllida Lloyd/*Duchess of Malfi* press clippings file. National Theatre Archives, London.
Foucault, Michel. (1995/1977). *Discipline and Punish: The Birth of the Prison*. Trans. Alan Sheridan. New York: Vintage.
Foxe, John. (1563). *Actes and Monuments*. London: John Day.
Freedman, Barbara. (1990). 'Frame-Up: Feminism, Psychoanalysis, Theatre.' *Performing Feminisms: Feminist Critical Theory and Theatre*. Ed. Sue Ellen Case. Baltimore, MD: Johns Hopkins University Press, pp. 54–76.
Freud, Sigmund. (1953–1974). 'The Dissolution of the Oedipus Complex.' 1924. *The Standard Edition of the Complete Psychological Works of Sigmund Freud*. Vol. 19. Ed. and trans. James Strachey. London: The Hogarth Press and the Institute of Psychoanalysis, pp. 171–9.
———. (1953–1974). 'On the Sexual Theories of Children.' 1908. *The Standard Edition of the Complete Psychological Works of Sigmund Freud*. Vol. 9. Ed. and trans. James Strachey. London: The Hogarth Press and the Institute of Psychoanalysis, pp. 205–26.
———. (1953–1974). 'Some Psychical Consequences of the Anatomical Distinction Between the Sexes.' 1925. *The Standard Edition of the Complete Psychological Works of Sigmund Freud*. Vol. 19. Ed. and trans. James Strachey. London: The Hogarth Press and the Institute of Psychoanalysis, pp. 241–58.
———. (1953–1974). 'Three Essays on the Theory of Sexuality.' 1905. *The Standard Edition of the Complete Psychological Works of Sigmund Freud*. Vol. 7. Ed. and trans. James Strachey. London: The Hogarth Press and the Institute of Psychoanalysis, pp. 123–245.
———. (1986). 'Femininity.' 1933. *The Essentials of Psychoanalysis*. Ed. Anna Freud. Trans. James Strachey. London: Penguin, pp. 412–32.

Frey, Christopher and Leanore Lieblein. (2004). '"My breasts sear'd": The Self-Starved Female Body and *A Woman Killed with Kindness*.' *Early Theatre*, 7.1: 45–66.
Friedman, Alice T. (1989). *House and Household in Elizabethan England: Wollaton Hall and the Willoughby Family*. Chicago: University of Chicago Press.
Gainor, J. Ellen. (2002). 'Rethinking Feminism, Stanislavsky, and Performance.' *Theatre Topics*, 12.2: 163–75.
Garber, Marjorie. (1996). 'The Insincerity of Women.' 1994. *Subject and Object in Renaissance Culture*. Ed. Margreta de Grazia, Maureen Quilligan, and Peter Stallybrass. Cambridge: Cambridge University Press, pp. 349–68.
Gardner, Lyn. (1987). Review of *Titus Andronicus*, by William Shakespeare. *Theatre Record*, 7–20 May, p. 634. (Rpt from *City Limits*, 28 May 1987.)
Garebian, Keith. (n.d.). Review of *The Duchess of Malfi*, by John Webster. *Stage and Page* (online). Peter Hinton/*Duchess of Malfi* clippings file. Stratford Shakespeare Festival Archives, Stratford, ON.
Giannachi, Gabriella, and Mary Luckhurst. (1999). Interview with Katie Mitchell. *On Directing: Interviews with Directors*. Ed. Gabriella Giannachi and Mary Luckhurst. London: Faber, pp. 95–102.
Glanvill. (1993). *The Treatise on the Laws and Customes of the Realm of England Commonly Called Glanvill*. Ed. and trans. G. D. G. Hall. Oxford: Clarendon.
Godfrey, Robyn. (2006). 'It was a dark and stormy night.' *Stratford City Gazette*, 16 June. Peter Hinton/*Duchess of Malfi* clippings file. Stratford Shakespeare Festival Archives, Stratford, ON.
Gordon, Giles. (1987). Review of *Titus Andronicus*, by William Shakespeare. *Theatre Record*, 7–20 May, pp. 634–5. (Rpt from *The London Daily News*, 13 May 1987.)
Gore-Langton, Robert. (2003). Review of *The Duchess of Malfi*, by John Webster. *Theatre Record*, 1–28 January, p. 76. (Rpt. from *The Express*, 31 January 2003.)
Gouge, William. (1976). *Of Domesticall Duties*. 1622. Amsterdam: Theatrum Orbis Terrarum; Norwood, NJ: Walter J. Johnson.
Gowing, Laura. (1996). *Domestic Dangers: Women, Words, and Sex in Early Modern London*. Oxford Studies in Social History. Oxford: Clarendon.
——. (2003). *Common Bodies: Women, Touch and Power in Seventeenth-Century England*. New Haven, CT, and London: Yale University Press.
Gravdal, Kathryn. (1991). *Ravishing Maidens: Writing Rape in Medieval French Literature and Law*. Philadelphia: University of Pennsylvania Press.
Graver, David. (1995). 'Violent Theatricality: Displayed Enactments of Aggression and Pain.' *Theatre Journal*, 47: 43–64.
Green, Douglas E. (1989). 'Interpreting "her martyr'd signs": Gender and Tragedy in *Titus Andronicus*.' *Shakespeare Quarterly*, 40.3: 317–26.
Green, Reina. (2003). '"Ears Prejudicate" in *Mariam* and *Duchess of Malfi*.' *SEL*, 43.2: 459–74.
Gross, Kenneth. (2001). *Shakespeare's Noise*. Chicago and London: University of Chicago Press.
Grosz, Elizabeth. (1995). 'Women, *Chora*, Dwelling.' *Space, Time, and Perversion*. New York and London: Routledge, pp. 11–24.

Gutierrez, Nancy A. (1989). 'The Irresolution of Melodrama: The Meaning of Adultery in *A Woman Killed With Kindness.*' *Exemplaria*, 1.2: 265–91.
——. (1994). 'Exorcism by Fasting in *A Woman Killed With Kindness*: A Paradigm of Puritan Resistance?' *Research Opportunities in Renaissance Drama*, 33.1–2: 43–62.
Haber, Judith. (1997). '"My Body Bestow upon My Women": The Space of the Feminine in *The Duchess of Malfi.*' *Renaissance Drama*, 28: 133–59.
——. (2003). '"I(t) could not choose but follow": Erotic Logic in *The Changeling.*' *Representations*, 81: 79–98.
Hanawalt, Barbara A. (1998). 'Whose Story Was This? Rape Narratives in Medieval English Courts.' *'Of Good and Ill Repute': Gender and Social Control in Medieval England.* Oxford: Oxford University Press, pp. 124–41.
Harvey, Elizabeth D. (2003a). 'The "Sense of All Senses."' *Sensible Flesh: On Touch in Early Modern Culture.* Ed. Elizabeth D. Harvey. Philadelphia: University of Pennsylvania Press, pp. 1–21.
——, ed. (2003b). *Sensible Flesh: On Touch in Early Modern Culture.* Philadelphia: University of Pennsylvania Press.
——. (2003c). 'The Touching Organ: Allegory, Anatomy, and the Renaissance Skin Envelope.' *Sensible Flesh: On Touch in Early Modern Culture.* Ed. Elizabeth D. Harvey. Philadelphia: University of Pennsylvania Press, pp. 81–102.
Heale, William. (1974). *An Apologie for Women.* 1609. Amsterdam: Theatrum Orbis Terrarum; Norwood, NJ: Walter J. Johnson.
Healy, Margaret. (2003). 'Anxious and Fatal Contacts: Taming the Contagious Touch.' *Sensible Flesh: On Touch in Early Modern Culture.* Ed. Elizabeth D. Harvey. Philadelphia: University of Pennsylvania Press, pp. 22–38.
Hedrick, Donald, and Bryan Reynolds. (2006). 'I Might Like You Better If We Slept Together: The Historical Drift of Place in *The Changeling.*' *Transversal Enterprises in the Drama of Shakespeare and His Contemporaries: Fugitive Explorations.* Basingstoke: Palgrave Macmillan, pp. 112–23.
Henderson, Andrea. (2000). 'Death on Stage, Death of the Stage: The Antitheatricality of *The Duchess of Malfi.*' 1990. *The Duchess of Malfi: Contemporary Critical Essays.* Ed. Dympna Callaghan. New Casebooks. Basingstoke: Palgrave Macmillan, pp. 61–79.
Herd, Juliet. (2003). 'Modern-day Malfi.' *Hello!*, 11 February. Phyllida Lloyd/ *Duchess of Malfi* press clippings file. National Theatre Archives, London.
Heywood, Thomas. (1950). *The Rape of Lucrece.* Ed. Alan Holaday. Urbana: University of Illinois Press.
——. (1985). *A Woman Killed With Kindness.* Ed. Brian W. M. Scobie. New Mermaids. London: A. & C. Black.
——. (1992). *A Woman Killed With Kindness.* Dir. Katie Mitchell. The Pit, London: Royal Shakespeare Company. First performance: April 1992. Archive Video Recording. The Shakespeare Centre Library, Stratford-upon-Avon.
Higgins, Lynn A., and Brenda R. Silver, eds. (1991). *Rape and Representation.* New York: Columbia University Press.
Hiley, Jim. (1988). Review of *Titus Andronicus*, by William Shakespeare. *Theatre Record*, 1–14 July, pp. 904–5. (Rpt. from *The Listener*, 14 July 1988.)
Hills, Helen. (2003). 'Introduction: Theorizing the Relationship Between Architecture and Gender in Early Modern Europe.' *Architecture and the*

Politics of Gender in Early Modern Europe. Ed. Helen Hills. Aldershot: Ashgate, pp. 3–22.

Hoile, Christopher. (n.d.). 'Oh, Direful Misprision!' *Stage Door*. Peter Hinton/ *Duchess of Malfi* clippings file. Stratford Shakespeare Festival Archives, Stratford, ON.

An Homily of the State of Matrimony. (1992). 1563. *Daughters, Wives, and Widows: Writings by Men about Women and Marriage in England, 1500–1640*. Ed. Joan Larsen Klein. Urbana and Chicago: University of Illinois Press, pp. 11–25.

Holden, Stephen. (1999). 'It's Sort of Family Dinner, Your Majesty.' Review of *Titus*, by Julie Taymor. *New York Times* (online), 24 December. Available at: <http://query.nytimes.com/gst/fullpage.html?res9C00E0D71539F937A15751C 1A96F958260&scp7&sqtaymor+titus&stnyt>.

Honess-Martin, Jacqui. (2007). Personal interview. 23 November.

Hopkins, D.J. (2008). *City/Stage/Globe: Performance and Space in Shakespeare's London*. New York: Routledge.

Hopkins, Lisa. (1997). 'Beguiling the Master of the Mystery: Form and Power in *The Changeling*.' *Medieval and Renaissance Drama in England*, 9: 149–61.

———. (2002). *The Female Hero in English Renaissance Tragedy*. Basingstoke: Palgrave Macmillan.

———. (2003a). ' "A tiger's heart wrapped in a player's hide": Julie Taymor's War Dances.' *Shakespeare Bulletin*, 21.3: 61–9.

———. (2003b). 'With the Skin Side Inside: The Interiors of *The Duchess of Malfi*.' *Privacy, Domesticity, and Women in Early Modern England*. Ed. Corinne S. Abate. Aldershot: Ashgate, pp. 21–30.

Howe, Eunice D. (2003). 'The Architecture of Institutionalism: Women's Space in Renaissance Hospitals.' *Architecture and the Politics of Gender in Early Modern Europe*. Ed. Helen Hills. Aldershot: Ashgate, pp. 63–82.

Hulse, S. Clark. (1979). 'Wresting the Alphabet: Oratory and Action in *Titus Andronicus*.' *Criticism*, 21.2: 106–18.

Hunt, Margaret. (1992). 'Wife Beating, Domesticity and Women's Independence in Eighteenth-Century London.' *Gender and History*, 4.1: 10–33.

Ingraham, Catherine. (1998). *Architecture and the Burdens of Linearity*. New Haven, CT, and London: Yale University Press.

Irvine, Susan. (2006). Review of *The Changeling*, by Thomas Middleton and William Rowley. *Theatre Record*, 7–20 May, p. 577. (Rpt. from *The Sunday Telegraph*, 21 May 2006.)

Jankowski, Theodora. (1992). *Women in Power in Early Modern England*. Urbana: University of Illinois Press.

Jardine, Lisa. (1989). *Still Harping on Daughters: Women and Drama in the Age of Shakespeare*, 2nd edn. New York: Columbia University Press.

———. (1996). 'Companionate Marriage Versus Male Friendship: Anxiety for the Lineal Family in Jacobean Drama.' *Reading Shakespeare Historically*. London: Routledge, pp. 114–31.

Johnson, Aidan. (2006). 'Stratford play a haunting vision of the macabre.' 6 September. Peter Hinton/*Duchess of Malfi* clippings file. Stratford Shakespeare Festival Archives, Stratford, ON.

Johnson, Lawrence B. (2006). 'Love's a family matter in "Duchess of Malfi" at Stratford fest.' *Detroit News*, 15 June. Peter Hinton/*Duchess of Malfi* clippings file. Stratford Shakespeare Festival Archives, Stratford, ON.

Joplin, Patricia Klindienst. (1991). 'The Voice of the Shuttle Is Ours.' *Rape and Representation*. Ed. Lynn A. Higgins and Brenda R. Silver. New York: Columbia University Press, pp. 35–64.

Jordan, Jan. (2004). *The Word of a Woman? Police, Rape and Belief*. Basingstoke: Palgrave Macmillan.

Jordan, J. T. (2006). 'Stark drama lacks hope, sense of subtlety.' *New Hamburg Independent*, 2 August, p. 13. Peter Hinton/*Duchess of Malfi* clippings file. Stratford Shakespeare Festival Archives, Stratford, ON.

Kahn, Coppélia. (1976). 'The Rape in Shakespeare's *Lucrece*.' *Shakespeare Studies*, 9: 45–72.

Kaplan, E. Ann. (1983). 'Is the Gaze Male?' *Powers of Desire: The Politics of Sexuality*. Ed. Ann Snitow, Christine Stansell, and Sharon Thompson. New York: Monthly Review Press, pp. 309–27.

Kaplan, Jon. (2006). 'This Duchess is dark.' *NOW*, 7–13 September. Peter Hinton/ *Duchess of Malfi* clippings file. Stratford Shakespeare Festival Archives, Stratford, ON.

Keidan, Lois. (2006). 'This Must be the Place: Thoughts on Place, Placelessness and Live Art since the 1980s.' *Performance and Place*. Ed. Leslie Hill and Helen Paris. Basingstoke: Palgrave Macmillan.

Kendall, Gillian Murray. (1989). '"Lend Me Thy Hand": Metaphor and Mayhem in *Titus Andronicus*.' *Shakespeare Quarterly*, 40.3: 299–316.

Kidnie, Margaret Jane, ed. (2002). *The Anatomie of Abuses*. By Phillip Stubbes. Tempe, AZ: Arizona Center for Medieval and Renaissance Studies/Renaissance English Text Society.

Kiefer, Frederick. (1986). 'Heywood as Moralist in *A Woman Killed With Kindness*.' *Medieval and Renaissance Drama in England*, 3: 83–98.

Kintz, Linda. (27 July 2007). Panel Participant. 'Regenerating Praxis: Roundtable on Empathy, Activism, and Performance in Times of Crisis.' ATHE 2007 Annual Conference. Sheraton New Orleans, New Orleans, LA.

Klein, Joan Larsen, ed. (1992). *Daughters, Wives, and Widows: Writings by Men about Women and Marriage in England, 1500–1640*. Urbana and Chicago: University of Illinois Press.

Knott, John R. (1993). *Discourses of Martyrdom in English Literature, 1563–1694*. Cambridge: Cambridge University Press.

Knowles, Richard Paul. (2004a). *Reading the Material Theatre*. Cambridge: Cambridge University Press.

——. (2004b). *Shakespeare and Canada: Essays on Production, Translation, and Adaptation*. Brussels and New York: Peter Lang.

Konradi, Amanda. (2007). *Taking the Stand: Rape Survivors and the Prosecution of Rapists*. Westport, CT: Praeger.

Krasner, David. (2006). 'Empathy and Theater.' *Staging Philosophy: Intersections of Theater, Performance, and Philosophy*. Ed. David Krasner and David Z. Saltz. Ann Arbor: University of Michigan Press, pp. 255–77.

Kristeva, Julia. (1986). 'The True-Real.' Trans. Seán Hand. *The Kristeva Reader*. Ed. Toril Moi. New York: Columbia University Press, pp. 216–37.

Kubiak, Anthony. (1991). *Stages of Terror: Terrorism, Ideology, and Coercion as Theatre History.* Bloomington and Indianapolis: Indiana University Press.

Lacan, Jacques. (1968). 'The Function of Language in Psychoanalysis.' 1956. *The Language of the Self: The Function of Language in Psychoanalysis.* Trans. Anthony Wilden. Baltimore, MD: Johns Hopkins University Press, pp. 1–87.

Lehmann, Courtney, Bryan Reynolds, and Lisa Starks. (2003). '"For Such a Sight Will Blind a Father's Eye": The Spectacle of Suffering in Taymor's *Titus*.' *Performing Transversally: Reimagining Shakespeare and the Critical Future.* Ed. Bryan Reynolds. Basingstoke: Palgrave Macmillan, pp. 215–43.

Letts, Quentin. (2006). 'Don't go changeling to try and please me.' Review of *The Changeling*, by Thomas Middleton and William Rowley. *Theatre Record*, 7–20 May, pp. 576–7. (Rpt. from *The Daily Mail*, 19 May 2006.)

Liebler, Naomi C. (2003). '"A Woman Dipped in Blood": The Violent Femmes of *The Maid's Tragedy* and *The Changeling*.' *Women, Violence, and English Renaissance Literature: Essays Honoring Paul Jorgensen.* Ed. Linda Woodbridge and Sharon Beehler. Tempe, AZ: Arizona Center for Medieval and Renaissance Studies, pp. 361–78.

Lucas, Valerie. (1990). 'Puritan Preaching and the Politics of the Family.' *The Renaissance Englishwoman in Print: Counterbalancing the Canon.* Ed. Anne M. Haselkorn and Betty S. Travitsky. Amherst: University of Massachusetts Press, pp. 224–40.

Luckyj, Christina. (2002). *A Moving Rhetoricke: Gender and Silence in Early Modern England.* Manchester: Manchester University Press.

Lutterbie, John. (2001). 'Phenomenology and the Dramaturgy of Space and Place.' *Journal of Dramatic Theory and Criticism*, 16.1: 123–30.

Macaulay, Alastair. (2003). 'Lloyd is no match for the Duchess.' *Theatre Record*, 1–28 January, p. 72. (Rpt. from *The Financial Times*, 30 January 2003.)

MacDonald, Joyce Green. (1993). 'Women and Theatrical Authority: Deborah Warner's *Titus Andronicus*.' *Cross-Cultural Performances: Differences in Women's Re-Visions of Shakespeare.* Ed. Marianne Novy. Urbana: University of Illinois Press, pp. 185–205.

Malcolmson, Cristina. (2001). '"As Tame as the Ladies": Politics and Gender in *The Changeling*.' 1990. *Revenge Tragedy.* Ed. Stevie Simkin. Basingstoke and New York: Palgrave Macmillan, pp. 142–62.

Manfull, Helen. (1997). *In Other Words: Women Directors Speak.* Lyme, NH: Smith & Kraus.

Margolin, Deb. (2008). '"To Speak is to Suffer" and Vice Versa.' *TDR*, 52.3: 95–7.

Marmion, Patrick. (1988). Review of *Titus Andronicus*, by William Shakespeare. *Theatre Record*, 1–14 July, p. 903. (Rpt from *What's On*, 13 July 1988.)

Marshall, Cynthia. (1991). '"I can interpret all her martyr'd signs": *Titus Andronicus*, Feminism, and the Limits of Interpretation.' *Sexuality and Politics in Renaissance Drama.* Ed. Carole Levin and Karen Robertson. Lewiston, NY: Edwin Mellen, pp. 193–213.

———. (2002). *The Shattering of the Self: Violence, Subjectivity, and Early Modern Texts.* Baltimore, MD, and London: Johns Hopkins University Press.

Mazzio, Carla. (2003). 'Acting with Tact: Touch and Theater in the Renaissance.' *Sensible Flesh: On Touch in Early Modern Culture.* Ed. Elizabeth D. Harvey. Philadelphia: University of Pennsylvania Press, pp. 159–86.

McAfee, Annalena. (1988). Review of *Titus Andronicus*, by William Shakespeare. *Theatre Record*, 1–14 July, p. 903. (Rpt from *The Evening Standard*, 5 July 1988.)
McAuley, Gay. (1999). *Space in Performance: Making Meaning in the Theatre*. Ann Arbor: University of Michigan Press.
McCandless, David. (2002). 'A Tale of Two Tit*uses*: Julie Taymor's Vision on Stage and Screen.' *Shakespeare Quarterly*, 53.4: 487–511.
McEwen, Indra Kagis. (2003). *Vitruvius: Writing the Body of Architecture*. Cambridge, MA: MIT Press.
McKinnie, Michael (2009). 'Performing the Civic Transnational: Cultural Production, Governance, and Citizenship in Contemporary London.' *Performance and the City*. Ed. D.J. Hopkins, Shelley Orr, and Kim Solga. Basingstoke: Palgrave, pp. 110–27.
McLuskie, Kathleen. (1989). *Renaissance Dramatists*. Atlantic Highlands, NJ: Humanities Press International.
——. (2000). 'Drama and Sexual Politics: The Case of Webster's Duchess.' 1985. *The Duchess of Malfi: Contemporary Critical Essays*. Ed. Dympna Callaghan. New Casebooks. Basingstoke: Palgrave Macmillan, pp. 104–21.
McLuskie, Kathleen, and Jennifer Uglow, eds. (1989). *The Duchess of Malfi*. Plays in Performance. Bristol: Bristol Classical Press.
McQuade, Paula. (2000). ' "A Labyrinth of Sin": Marriage and Moral Capacity in Thomas Heywood's *A Woman Killed With Kindness*.' *Modern Philology*, 98.2: 231–50.
Mendelson, Sara, and Patricia Crawford. (1998). *Women in Early Modern England: 1550–1720*. Oxford: Clarendon.
Middleton, Thomas, and William Rowley. (2006). *The Changeling*. Dir. and design Declan Donnellan and Nick Ormerod. The Barbican Centre, London: Cheek by Jowl. First performance: May 2006. Various European tour venues March – August 2006. Archive DVD recordings. Cheek by Jowl, the Barbican Centre, London.
——. (2006). *The Changeling*. Ed. Michael Neill. New Mermaids. London: A & C Black.
Miller, Graeme. (2006). 'Through the Wrong End of the Telescope.' *Performance and Place*. Ed. Leslie Hill and Helen Paris. Basingstoke: Palgrave Macmillan, pp. 104–12.
Morley, Sheridan. (1988). Review of *Titus Andronicus*, by William Shakespeare. *Theatre Record*, 1–14 July, p. 901. (Rpt from *Punch*, 22 July 1988.)
Morris, Sylvia. (2006). Personal interview. 12 July.
Mullaney, Steven. (1988). *The Place of the Stage: License, Play, and Power in Renaissance England*. Chicago: University of Chicago Press.
Mulvey, Laura. (1975). 'Visual Pleasure and Narrative Cinema.' *Screen*, 16.3: 6–18.
Nathan, John. (2003). 'Pass the Duchess.' *Theatre Record*, 1–28 January, p. 71. (Rpt. from *The Jewish Chronicle*, 31 January 2003.)
Neill, Michael. (2005). ' "A Woman's Service": Gender, Subordination, and the Erotics of Rank in the Drama of Shakespeare and His Contemporaries.' *The Shakespearean International Yearbook*, 5: 127–44.
——. (2006). Introduction. *The Changeling*. By Thomas Middleton and William Rowley. Ed. Michael Neill. New Mermaids. London: A & C Black, pp. vii–xlv.
Nightingale, Benedict. (2003). 'Black suits don't fit in classic tragedy.' *Theatre Record*, 1–28 January, p. 71. (Rpt. from *The Times*, 29 January 2003.)

O'Connell, Michael. (2000). *The Idolatrous Eye: Iconoclasm and Theater in Early-Modern England*. New York and Oxford: Oxford University Press.
Oliver, Kelly. (1998). *Subjectivity Without Subjects: From Abject Fathers to Desiring Mothers*. Lanham, MD: Rowman & Littlefield.
——. (2001). *Witnessing: Beyond Recognition*. Minneapolis: University of Minnesota Press.
Orlin, Lena Cowen. (1994). *Private Matters and Public Culture in Post-Reformation England*. Ithaca, NY, and London: Cornell University Press.
Orrell, John. (1985). *The Theatres of Inigo Jones and John Webb*. Cambridge and New York: Cambridge University Press.
Ouzounian, Richard. (2006). 'Gloominess made much gloomier.' *The Toronto Star*, 3 June, p. A 25.
Owens, Margaret E. (2005). 'The Revenge of the Martyred Body: R.B.'s Appius and Virginia.' *Stages of Dismemberment: The Fragmented Body in Late Medieval and Early Modern Drama*. Newark: University of Delaware Press, 84–114.
Oxford English Dictionary. (2007). s.v. 'Complain' (online). 9 July 2007. Available at: <http://dictionary.oed.com>.
——. (2007). s.v. 'Kindness' (online). January 2007. Available at: <http://dictionary.oed.com>.
Palladio, Andrea. (1997). *The Four Books on Architecture*. Trans. Robert Tavernor and Richard Schofield. Cambridge, MA: MIT Press.
Palmer, D. J. (1972). 'The Unspeakable in Pursuit of the Uneatable: Language and Action in *Titus Andronicus*.' *Critical Quarterly*, 14.4: 320–39.
Panek, Jennifer. (1994). 'Punishing Adultery in *A Woman Killed With Kindness*.' *SEL*, 34.2: 357–78.
Payne, Alina A. (1999). *The Architectural Treatise in the Italian Renaissance: Architectural Invention, Ornament, and Literary Culture*. Cambridge: Cambridge University Press.
Peter, John. (2003). Review of *The Duchess of Malfi*, by John Webster. *Theatre Record*, 1–28 January, p. 75. (Rpt. from *The Sunday Times*, 2 February 2003.)
——. (2006). Review of *The Changeling*, by Thomas Middleton and William Rowley. *Theatre Record*, 7–20 May, p. 577. (Rpt. from *The Sunday Times*, 21 May 2006.)
Peters, Christine. (2003). *Patterns of Piety: Women, Gender and Religion in Late Medieval and Reformation England*. Cambridge Studies in Early Modern British History. Cambridge: Cambridge University Press.
Peterson, Kaara L. (2004). 'Shakespearean Revivifications: Early Modern Undead.' *Shakespeare Studies*, 32: 240–66.
Phelan, Peggy. (1993). *Unmarked: The Politics of Performance*. London and New York: Routledge.
——. (1997). *Mourning Sex: Performing Public Memories*. London: Routledge.
Phillippy, Patricia. (2002). *Women, Death and Literature in Post-Reformation England*. Cambridge: Cambridge University Press.
Phillips, Kim M. (2000). 'Written on the Body: Reading Rape from the Twelfth to Fifteenth Centuries.' *Medieval Women and the Law*. Ed. Noël James Menuge. Woodbridge, Suffolk: Boydell, pp. 125–44.
Portman, Jamie. (2006). 'Stratford's Duchess more style than substance.' CanWest News Service, 5 June. Peter Hinton/*Duchess of Malfi* clippings file. Stratford Shakespeare Festival Archives, Stratford, ON.

Post, J. B. (1978). 'Ravishment of Women and the Statutes of Westminster.' *Legal Records and the Historian*. Ed. J. H. Baker. London: Royal Historical Society, pp. 150–64.

Ramsden, Timothy. (2006). Review of *The Changeling*, by Thomas Middleton and William Rowley. *Reviews Gate* (online), 8 May. 19 November 2007. Available at: <http://www.reviewsgate.com/index.php?nameNews&filearticle&sid2858>.

Rayburn, Corey. (2006). 'To Catch a Sex Thief: The Burden of Performance in Rape and Sexual Assault Trials.' *Columbia Journal of Gender and Law* (online), 15.2. Available at: <http://proquest.umi.com>.

Rayner, Alice. (2006). *Ghosts: Death's Double and the Phenomena of Theatre*. Minneapolis: University of Minnesota Press.

Reid, Robert. (2006). 'The Duchess of Malfi not for the faint of heart.' *Kitchener-Waterloo Record*, 5 June. Peter Hinton/*Duchess of Malfi* clippings file. Stratford Shakespeare Festival Archives, Stratford, ON.

Ridout, Nicholas. (2006). *Stage Fright, Animals, and Other Theatrical Problems*. Cambridge: Cambridge University Press.

Roach, Joseph. (1996). *Cities of the Dead: Circum-Atlantic Performance*. New York: Columbia University Press.

Roberts, Anna, ed. (1998). *Violence Against Women in Medieval Texts*. Gainesville: University Press of Florida.

Robertson, Karen. (2001). 'Rape and the Appropriation of Progne's Revenge in Shakespeare's *Titus Andronicus*, Or "Who Cooks the Thyestean Banquet?"' *Representing Rape in Medieval and Early Modern Literature*. Ed. Elizabeth Robertson and Christine M. Rose. New York: Palgrave Macmillan, pp. 213–37.

Román, David. (1998). *Acts of Intervention: Performance, Gay Culture, and AIDS*. Indianapolis: Indiana University Press.

——. (2005). *Performance in America: Contemporary US Culture and the Performing Arts*. Durham, NC, and London: Duke University Press.

Rose, Mary Beth. (1988). *The Expense of Spirit: Love and Sexuality in English Renaissance Drama*. Ithaca, NY: Cornell University Press.

Rossini, Manuela S. (1998). 'The New Domestic Ethic in English Renaissance Drama: Thomas Heywood's *A Woman Killed With Kindness* (1603).' *A Woman's Place: Women, Domesticity and Private Life*. Ed. Annabelle Despard. Kristiansand: Agder College, pp. 106–17.

Rowe, Katherine A. (1994). 'Dismembering and Forgetting in *Titus Andronicus*.' *Shakespeare Quarterly*, 45.3: 279–303.

Schneider, Rebecca. (2001). 'Performance Remains.' *Performance Research*, 6.2: 100–8.

——. (2009). 'Patricidal Memory and the Passerby.' *Performance and the City*. Ed. D.J. Hopkins, Shelley Orr, and Kim Solga. Basingstoke: Palgrave Macmillan.

Scott, Adam. (n.d.). Review of *The Duchess of Malfi*, by John Webster. *The Treatment* (?). Phyllida Lloyd/*Duchess of Malfi* press clippings file. National Theatre Archives, London.

Scowcroft, Philippa. (2006). Review of *The Duchess of Malfi*, by John Webster. 22 June. Peter Hinton/*Duchess of Malfi* clippings file. Stratford Shakespeare Festival Archives, Stratford, ON.

Shakespeare, William. (1969). *Othello*. Ed. Gerald Eades Bentley. *The Complete Works*. New York: Viking/Penguin, pp. 1018–59.

——. (1987). *Titus Andronicus*. Dir. Deborah Warner. Swan Theatre in Stratford-upon-Avon: Royal Shakespeare Company. First performance: August 1987. Archive Video Recording. The Shakespeare Centre Library, Stratford-upon-Avon.

——. (1995). *Titus Andronicus*. Ed. Jonathan Bate. London: Routledge.

——. (1996). *The Taming of the Shrew*. Ed. David Bevington. *The Taming of the Shrew: Texts and Contexts*. Ed. Frances E Dolan. Boston, MA: Bedford, pp. 41–139.

Sheffield, Graham. (2006). 'Welcome to the Barbican.' Programme. *The Changeling* and *Twelfth Night*. The Barbican Centre/Cheek by Jowl.

Shenton, Mark. (2006). Review of *The Changeling*, by Thomas Middleton and William Rowley. *Theatre Record*, 7–20 May, p. 578. (Rpt. from *What's On*, 26 May 2006.)

Shevtsova, Maria. (2006). 'On Directing: A Conversation with Katie Mitchell.' *New Theatre Quarterly*, 22.1: 3–18.

Shildrick, Margrit. (2002). *Embodying the Monster: Encounters with the Vulnerable Self*. London: Sage.

Shuger, Debora Kuller. (1994). *The Renaissance Bible: Scholarship, Sacrifice, and Subjectivity*. Berkeley: University of California Press.

Shuttleworth, Ian. (2006). Review of *The Changeling*, by Thomas Middleton and William Rowley. *Theatre Record*, 7–20 May, p. 576. (Rpt. from *The Financial Times*, 17 May 2006.)

Simon, Roger I. (2005). *The Touch of the Past: Remembrance, Learning, and Ethics*. New York: Palgrave Macmillan.

Simpson, Herbert. (2006). Review of *The Duchess of Malfi*, by John Webster. *Total Theater Online* (online), June. Peter Hinton/*Duchess of Malfi* clippings file. Stratford Shakespeare Festival Archives, Stratford, ON.

Smart, Carol. (1989). *Feminism and the Power of Law*. London and New York: Routledge.

Smith, Gary. (n.d.). 'Splendour spoils a good play.' *Hamilton Spectator*. Peter Hinton/*Duchess of Malfi* clippings file. Stratford Shakespeare Festival Archives, Stratford, ON.

Sommerville, Margaret R. (1995). *Sex and Subjection: Attitudes to Women in Early-Modern Society*. London: Arnold.

Spencer, Charles. (1988). Review of *Titus Andronicus*, by William Shakespeare. *Theatre Record*, 1–14 July, pp. 905–6. (Rpt from *The Daily Telegraph*, 6 July 1988.)

——. (2003). 'Very rocky horror show.' *Theatre Record*, 1–28 January, pp. 72–3. (Rpt. from *The Daily Telegraph*, 31 January 2003.)

——. (2006). 'An absolute cracker cloaked in silliness.' Review of *The Changeling*, by Thomas Middleton and William Rowley. *Theatre Record*, 7–20 May, p. 576. (Rpt. from *The Daily Telegraph*, 17 May 2006.)

Stage Beauty. (2004). DVD. Dir. Richard Eyre. London, Qwerty Films.

Stimpson, Catharine R. (1980). 'Shakespeare and the Soil of Rape.' *The Woman's Part: Feminist Criticism of Shakespeare*. Ed. Carolyn Ruth Swift Lenz, Gayle Greene, and Carol Thomas Neely. Urbana: University of Illinois Press, pp. 56–64.

Stockholder, Kay. (1996). 'The Aristocratic Woman as Scapegoat: Romantic Love and Class Antagonism in *The Spanish Tragedy*, *The Duchess of Malfi*, and

The Changeling.' The Elizabethan Theatre. Vol. 14. Ed. A. L. Magnusson and C. E. McGee. Toronto: Meany, pp. 127–51.

Stubbes, Philip. (1992). *A Crystal Glass for Christian Women, Containing a Most Excellent Discourse of the Godly Life and Christian Death of Mistress Katherine Stubbes.* 1591. *Daughters, Wives, and Widows: Writings by Men about Women and Marriage in England, 1500–1640.* Ed. Joan Larsen Klein. Urbana and Chicago: University of Illinois Press, pp. 139–49.

Sugimura, N. K. (2006). 'Changelings and *The Changeling.*' *Essays in Criticism*, 56.3: 241–63.

Taylor, Diana. (1997). *Disappearing Acts: Spectacles of Gender and Nationalism in Argentina's 'Dirty War'.* Durham, NC, and London: Duke University Press.

——. (2003). *The Archive and the Repertoire: Performing Cultural Memory in the Americas.* Durham, NC: Duke University Press.

T.E. (1979). *The Lawes Resolutions of Womens Rights.* 1632. Amsterdam: Theatrum Orbis Terrarum; Norwood, NJ: Walter J. Johnson.

Tinker, Jack. (1988). Review of *Titus Andronicus*, by William Shakespeare. *Theatre Record*, 1–14 July, p. 902. (Rpt from *The Daily Mail*, 20 July 1988.)

Titus. (2000). DVD. Dir. Julie Taymor. Los Angeles: Twentieth-Century Fox.

Tompkins, Joanne. (2006). *Unsettling Space: Contestations in Contemporary Australian Theatre.* Basingstoke: Palgrave Macmillan.

Tricomi, Albert H. (1974). 'The Aesthetics of Mutilation in *Titus Andronicus.*' *Shakespeare Survey*, 27: 11–9.

Vitruvius. (1960). *The Ten Books on Architecture.* Trans. Morris Hicky Morgan. New York: Dover.

Walker, Garthine. (1998). 'Rereading Rape and Sexual Violence in Early Modern England.' *Gender and History*, 10.1: 1–25.

Wall, Wendy. (1993). *The Imprint of Gender: Authorship and Publication in the English Renaissance.* Ithaca, NY, and London: Cornell University Press.

——. (2006). 'Just a Spoonful of Sugar: Syrup and Domesticity in Early Modern England.' *Modern Philology*, 104.2: 149–72.

Wayne, Valerie, ed. (1992). *The Flower of Friendship: A Renaissance Dialogue Contesting Marriage.* By Edmund Tilney. Ithaca, NY, and London: Cornell University Press.

Webster, John. (1997). *The Duchess of Malfi.* Ed. John Russell Brown. Manchester: Manchester University Press.

——. (2003). *The Duchess of Malfi.* Dir. Phyllida Lloyd. Lyttelton Theatre, London: Royal National Theatre. First performance: January 2003. Archive Video Recording. National Theatre Archives, London.

——. (2006). *The Duchess of Malfi.* Dir. Peter Hinton. Tom Patterson Theatre, Stratford, ON: Stratford Festival of Canada. First performance: May 2006.

——. (2009). The Duchess of Malfi. Prompt script. Stratford Shakespeare Festival Archives. Stratford, ON.

Wentworth, Michael. (1990). 'Thomas Heywood's *A Woman Killed With Kindness* as Domestic Morality.' *Traditions and Innovations: Essays on British Literature of the Middle Ages and the Renaissance.* Ed. David G. Allen and Robert A. White. Newark: University of Delaware Press; London and Toronto: Associated University Press, pp. 150–62.

Whately, William. (1623). *A Bride-Bush, Or, A Direction for Married Persons*, 2nd edn. London: Printed by Bernard Alsop for Benjamin Fisher.

Whately, William. (1976). *A Care-Cloth, Or, The Cumbers and Troubles of Marriage*. 1624. Amsterdam: Theatrum Orbis Terrarum; Norwood, NJ: Walter J. Johnson.
Whigham, Frank. (1996). *Seizures of the Will in Early Modern English Drama*. Cambridge: Cambridge University Press.
Wigley, Mark. (1992). 'Untitled: The Housing of Gender.' *Sexuality and Space*. Ed. Beatriz Colomina. Princeton, NJ: Princeton Architectural Press, pp. 327–89.
Wiles, David. (2003). *A Short History of Western Performance Space*. Cambridge: Cambridge University Press.
Williams, Carolyn D. (1993). '"Silence, like a Lucrece knife": Shakespeare and the Meanings of Rape.' *Yearbook of English Studies*, 23: 93–110.
Williams, Linda. (1989). *Hard Core: Power, Pleasure, and the 'Frenzy of the Visible'*. Berkeley: University of California Press.
Williams, Olivia. (8 March 2006). Interview with Domenic Cavendish (online). 27 November 2007. Podcast available at: <http://www.cheekbyjowl.com/productions/thechangeling/podcast.html>.
———. (4 May 2006). 'I'm in rehearsal, and for the first time since I went into labour, I am afraid.' Tour Diary. *The Independent: Extra*, pp. 12–13.
Wolf, Matt. (2003). Review of *The Duchess of Malfi*, by John Webster. *Variety*, 3 February. Phyllida Lloyd/*Duchess of Malfi* press clippings file. National Theatre Archives, London.
Wolska, Aleksandra. (2005). 'Rabbits, Machines, and the Ontology of Performance.' *Theatre Journal*, 57: 83–95.
Woodbridge, Linda. (2003). Introduction. *Women, Violence, and English Renaissance Literature: Essays Honoring Paul Jorgensen*. Ed. Linda Woodbridge and Sharon Beehler. Tempe, AZ: Arizona Center for Medieval and Renaissance Studies, pp. xi–xlix.
Wynne-Davies, Marion. (1991). '"The Swallowing Womb": Consumed and Consuming Women in *Titus Andronicus*.' *The Matter of Difference: Materialist Feminist Criticism of Shakespeare*. Ed. Valerie Wayne. Ithaca, NY: Cornell University Press, pp. 129–51.
Zimmerman, Susan. (2005). *The Early Modern Corpse and Shakespeare's Theatre*. Edinburgh: Edinburgh University Press.

Index

adultery 66, 78, 142, 160
Aebischer, Pascale 43–5, 50–1, 53, 57, 59, 181 n.5
Agustí, Clara Escoda 57
Alberti, Leon Battista 151, 154–5, 158
Almagro Festival 186 n.1
al-Solaylee, Kamal 129
Amster, Mara 157
Amussen, Susan Dwyer 9–10, 68, 70, 182 n.3
Anderson, Christy 155
Annan, Jotham 143
architecture 27, 141–2, 144–5, 147, 151–6, 175
 closet 27, 143–4, 150–1, 156–62, 167–8, 170, 175
 of feminist performance 27, 145, 186 n.6
 fortress 144, 150–2, 156, 174
Askew, Anne 109–10
audience, 2, 7, 13, 17–19, 27–31, 41, 46–7, 51–60, 82, 87–8, 91, 95, 97, 99–100, 103, 112, 117–20, 126–50, 156, 160–3, 167–79; 184 n.2, 184 n.9, 185 n.2, 186 n.3
 reception desire 141–2, 145, 147, 168–9
 and witness (*see* witness)

Baines, Barbara J. 7, 8, 181 n.5, 186 n.3
Bal, Mieke 32
Bamford, Karen 8, 148, 180 n.2–3, 181 n.2, 181 n.5
Barbican Centre 27, 147, 162–3, 165, 168, 171–5, 185 n.1, 186 n.6
Barbican Pit 54
Barker, Roberta 2, 5, 28, 53, 59, 82, 87–8, 93, 111, 146–7, 184 n.6
Bashar, Nazife 8, 35
Bassett, Kate 174
Bassnett, Madeline 157

Bawcutt, N.W. 148
Behling, Laura L. 111
Belsey, Catherine 3–4, 56, 110
Benedict, David 167, 186 n.8
Bennett, Lyn 78
Bennett, Susan 2, 5
Berkowitz, Gerald 173
Bevington, David 104
Billington, Michael 54–5, 136, 185 n.13
Blair, Rhonda 183 n.13
Blau, Herbert 2, 4–5, 102, 180 n.4
Bloom, Gina 102, 115, 184 n.1
Bloomer, Jennifer 141, 144, 186 n.5
Boehrer, Bruce 158
Bonneville, Hugh 176
Bott, Robin L. 181 n.8
Bracton, Henri de (*Bracton treatise*) 34, 36–40, 71, 181 n.4
Brecht, Bertolt 88, 130, 134–6
Brechtian 53, 60, 86, 88, 130–1, 136, 162
Breuer, Josef, 89–90
Brison, Susan J. 33, 40
Bromley, Laura 78
Bronfen, Elizabeth 180
Brook, Peter 43, 60
Brown, Georgina 173, 185 n.13
Burks, Deborah 8, 34–5, 64, 142, 149, 157, 180 n.3, 181 n.5
Butler, Judith 94–5
Bynum, Caroline Walker 183 n.10

Callaghan, Dympna 184 n.1
Campion, Joyce 5–7, 17, 118, 120, 122–6, 127–31, 136, 140, 178
Carlson, Marvin 5, 145–6
Carty, Shane 117–18, 121, 124
Casey, Edward 145
Catty, Jocelyn 30–1
Cavendish, Domenic 163–4
Chakravorty, Swapan 148

Changeling, The 11, 27, 31, 141–53, 145–75, 156, 163, 179, 180 n.3, 186 n.4
 see also Cheek By Jowl; Declan Donnellan and Nick Ormerod; Thomas Middleton and William Rowley
chastity 34, 55, 60, 148–9, 151–2, 180 n.3
Chaudhuri Una, 150
Chaytor, Miranda 9, 35
Cheek by Jowl 27, 143–7, 159–63, 166, 171–5, 179, 185 n.1, 186 n.7
Clapp, Susannah 162
Clarke, Gregory 167
Cleaver, Robert (*A Godlie Forme of Householde Government*) 10, 66, 71, 73–4, 78
Coats, Catharine Randall 38, 108
Coddon, Karin 112
Collings, David 161, 164
Colomina, Beatriz 186
Comensoli, Viviana 75
companionism 75–7, 82, 183 n.6
Condlln, Laura 121, 124, 126
Correll, Barbara 184 n.1
Coulborn, John 130, 184 n.9
Coveney, Michael 136, 173, 185 n.13
Cox, Brian 51–5
Cranitch, Lorcan 131–5, 138
Crawford, Patricia 10, 69, 182 n.5
Creaser, Sharon 78
Crudup, Billy 176
Cumyn, Steve 117
Cushman, Robert 118

Daalder, Joost 104, 185 n.2
Daileader, Celia 104, 111, 184 n.1
Danes, Claire 176
Daniels, Ben 95
Davies, Kathleen 75, 77, 183 n.6
Davis, Lloyd 68–9
Davis, Tracy 2
De Jongh, Nicolas 173–4, 185 n.13
De Lauretis, Teresa 20, 180 n.1
Desmet, Christy 103

Detmer, Emily 67–8, 70, 72, 76, 182 n.4
Emily Detmer-Goebel 46, 181 n.7
Diamond, Elin 20, 41, 47, 67, 90, 102
Diehl, Huston 64, 112, 184 n.2
Dolan, Frances 9, 68, 72, 76, 107, 109–10, 182 n.4
Dolan, Jill 13, 28, 180 n.4
domestic violence 2, 7, 9, 11, 25, 63, 65, 74, 76, 83
 see also violence
Donnellan, Declan 143, 145, 162–72, 174, 178–9
Duchêne, Kate 95
Duchess of Malfi, The 4–7, 11, 17, 22, 27, 98–106, 111–141, 148, 153, 178, 184 n.1–3, 184 n.5, 184 n.8, 185 n.12–13
 see also Peter Hinton; Phyllida Lloyd; John Webster

Edwardes, Jane 166, 185 n.13
Edwards, Charles 131
Edwards, Christopher 54
Enders, Jody 21, 29–31, 58
Estrich, Susan 180 n.1
Euripides (*Iphigenia at Aulis*) 95
 see also Katie Mitchell
Everett, Rupert 176
execution *see* violence

fasting *see* violence
Fawcett, Mary Laughlin 181 n.6
feminism 1, 4, 11, 19–20, 22, 27–8, 34, 42–5, 47, 50–2, 56, 61, 74–6, 101, 104, 107, 117, 128, 130, 142, 144, 153, 165, 179, 182 n.13, 184 n.1, 186 n.5
feminist performance 3–4, 13, 19–21, 27, 32, 42, 56, 103, 139, 145, 179, 182 n.13, 183 n.13
feminist spectators 117, 179
and film 19–20, 180 n.5, 182 n.13
and realism 32, 183. n.13
feminine 59–60, 111, 116, 153, 155, 158
Festa, Angelika 23–6

First (and Second) Statutes of Westminster 8, 34, 39
Flather, Amanda 152
Fletcher, Anthony 9, 180 n.2
Forte, Jeanie 20–1
Foss, Roger 136, 185
Foucault, Michel 43, 67, 150
Foxe, John (*Actes and Monuments*) 108–10
Fraser, Laura 58–60, 62, 178, 182 n.13
Freedman, Barbara 180 n.5
Freud, Sigmund 14, 47–8, 89–90, 167, 181 n.9, 183 n.13
Freudian, 163, 170
pre-Freudian 147
pseudo-Freudian 163
Frey, Christopher, and Leanore Lieblein, 84, 183 n.10
Friedman, Alice 152–3, 157

Gainor, J. Ellen 183 n.13
Garber, Marjorie 157, 159
Gardner, Lyn 54
Garebian, Keith 129
gaze, the *see* violence
ghost *and* ghosting 5–7, 15–17, 22, 27–8, 53, 71, 73, 89, 98, 127–8, 170, 179
The Changeling 142
The Duchess of Malfi 4–7, 17, 22, 27, 98, 101–3, 115, 117–20, 122, 124–6, 127–9, 142, 184 n.8
and audience 4–7, 27, 118–22, 179
and performance theory 5–7, 15–17
Titus Andronicus 53
A Woman Killed with Kindness 89
Giannachi, Gabriella and Mary Luckhurst 88
Glanvill treatise 8, 34, 36–9, 71, 181 n.4
Godfrey, Robyn 184 n.9
Gordon, Giles 54
Gordon, Leah 87
Gore-Langton, Robert 185 n.185
Gouge, William (*Of Domesticall Duties*), 10, 66, 73–8, 83, 183 n.7

Gowing, Laura 9–10, 35–6, 69–70, 75–7, 113–14, 116–17, 122–3, 181 n.3, 182 n.5
Graver, David 19, 23
Green, Douglas 181 n.8
Green, Reina 184 n.1
Greenwood, Judith 167
Grey, Lady Jane 184 n.4
Gross, Kenneth 82
Grosz, Elizabeth 186 n.5
Gutierrez, Nancy 67, 78, 183 n.10

Haber, Judith, 111, 116, 148, 184 n.1
Hanawalt, Barbara 9, 37, 40
Hannan, Peter 118, 126
Harvey, Elizabeth 114–15
Heale, William (*An Apologie for Women*), 9, 68–70
Healy, Margaret 114
Heatherington, Tom 150
Henderson, *And*rea 112
Heywood, Thomas 11, 26–7, 30, 66–7, 78–80, 99–100, 183 n.9
see A Woman Killed with Kindness
Hiddleston, Tom 143, 160–1, 165, 186 n.6
Higgins, Lynn and Brenda Silver 180 n.1
Hiley, Jim 54
Hills, Helen 152
Hinton, Peter 4, 6, 17, 27, 103, 117–20, 122–4, 126, 128–31, 136, 139, 178–9, 184 n.8, 185 n.11, 185 n.13
see The Duchess of Malfi
Hoile, Chistopher 129, 184 n.9
Holden, Stephen 182
Homily on the State of Matrimony, An 10, 72
Hopkins, D.J. 160
Hopkins, Lisa 57–8, 104, 148, 157–8, 165, 182 n.15
Hou, David, 121, 124
Howard, Frances 148
Howe, Eunice 152
Hulse, S. Clark 48, 181 n.6
Hunt, Margaret 69
Hutchinson, Derek 51

Ibbotson, Piers 51
invisibility 4, 5, 11, 15–16, 39, 40, 95, 111, 133, 139, 124, 142–5, 147, 150, 156, 162, 174, 177
   invisible act 16–19, 26–8, 32, 66, 101–2, 141
   silence 12, 14, 17, 44, 58, 120
Irigaray, Luce 41, 47
Irvine, Susan 163, 167

Jankowski, Theodora 103, 111, 184 n.1
Jardine, Lisa 103, 144, 152 157
Johnson, Aiden 130
Johnson, Lawrence 130
Jones, Inigo 155
Jones, Osheen 57
Jonson, Ben, 2, 148
Joplin, Patricia Klindienst 180 n.1, 181 n.5
Jordan, Jan 40, 130

Kahn, Coppélia 8, 181 n.5
Kaplan, E. Ann 180 n.5
Kaplan, Jon 130
Keen, Will 131–2, 143, 160, 167–70
Keidan, Lois 172
Kendall, Gillian Murray 181 n.6
Kidnie, Margaret Jane 106, 182 n.1
Kiefer, Frederick 78
Kintz, Linda 136
Klein, Joan Larson 64, 106
Knott, John 64, 109, 113,
Knowles, Richard Paul 185 n.10
Konradi, Amanda 32–3, 181 n.11
Krasner, David 127, 130, 135
Kristeva, Julia 21, 47
Kubiak, Anthony 21–3,

Lacan, Jacques 14, 125–6
   Lacanian psychoanalysis 13
   The Real 14, 21, 57, 125, 176
   see also psychoanalysis
Lange, Jessica 58
Lawes Resolutions of Womens Rights, The 36–9, 46, 49, 68, 71, 186 n.3
   T.E. 9, 68
Lefebvre, Henri 145

Lehmann, Courtney, Bryan Reynolds, and Lisa Starks 58
Leigh, Vivian 43, 60
Letts, Quentin 173
Levinas, Emmanuel 94–5
Liebler, Naomi 151, 157
Lloyd, Phyllida 27, 103, 130–1, 132, 134–40, 161, 178–9, 185 n.11–13
   see The Duchess of Malfi
Lucas, Valerie 75
Luckyj, Christina 114
Luhrmann, Baz (Romeo + Juliet) 57
Lutterbie, John 145

Macaulay, Alastair 185 n.13
MacDonald, Joyce Green, 50, 53–4
madness 43, 51, 113, 124, 128, 133–4, 137–8, 163, 166–7
   madhouse 27, 166–7
   madmen 104, 112, 124, 133, 137
Malcolmson, Cristina 148
Maloney, Michael 86–7, 89–94, 178
Manfull, Helen 88
Margolin, Deb 12
Marmion, Patrick 54
Marshall, Cynthia 33, 44, 108, 181 n.8, 184 n.5
martyr, 44, 54, 64, 74, 77, 81–2, 94, 98–101, 103–6, 108–13, 116, 184. n.3–5
   martyrdom 77, 98–101, 104–6, 108–14, 116, 126–7, 133–4
   martyrology 99, 106–8, 113, 117, 120, 125–6, 133, 139, 178
   script 27, 100–1, 106–7, 116, 127
Mazzio, Carla 114
McAfee, Annalena 54
McAuley, Gay 145
McCabe, Richard 51
McCandless, David 60
McCarroll, Luke 129
McEwen, Indra Kagis 153
McKinnie, Michael 172
McLuskie, Kathleen 75, 103, 107
   and Jennifer Uglow 184 n.8
McQuade, Paula 78
McTeer, Janet 131–6, 138, 161, 178, 185 n.12
Melichar, Alena 132, 138

Mendelson, Sarah   10, 69, 182 n.5
metatheatrical   29–30, 32, 39–43,
   46–7, 50–1, 55–6, 59, 61–3, 122,
   150, 163, 166, 178
Meyers, Jonathan Rhys,   58
Middleton, Thomas   2, 11, 130, 141,
   144, 151, 156, 164, 167, 174, 180
   n.3
   see *The Changeling*   180 n.1, 186 n.4
   The Second Maiden's Tragedy   180 n.2
   Women Beware Women   31, 148
Miller, Graeme   160
Mitchell, Katie   26, 67, 85–98, 126,
   131, 133, 168, 178–9, 183 n.11,
   13, 184 n.9
   Iphigenia at Aulis   95–6, 100,
   183 n.11
   see *A Woman Killed with Kindness*
Morahan, Hattie   95–6, 102, 133
Morley, Sheridan   54
Morris, Sylvia   183 n.2
Mullaney, Steven   147
Mulvey, Laura   19
   see violence *and* the gaze

Nathan, John   137
naturalism   26, 87–90, 96, 126
Neill, Michael   185 n.2
Newman, Karen   110
Nicol, David   146–7
Nightingale, Benedict   185 n.13

O'Connell, Michael   38
Oliver, Kelly   17–18, 63, 102, 117,
   136, 178
Olivier, Laurence   43, 60
Orlin, Lena Cowen   78–9, 151, 157
Ormerod, Nick   143, 145, 160,
   162–72, 174–5, 178–9, 186 n.7
Orrell, John,   144
Other Place, The (Stratford-upon-
   Avon)   183 n.11–12
Ouzounian Richard,   129,
Owens, Margaret   98, 104, 184 n.4

pain   20–1, 23, 31, 33, 39, 45, 47,
   50–6, 59, 64–7, 83, 85–91, 93, 99,
   103, 110, 113, 122–7, 137, 169,
   184 n.5

audience *and*   129, 162
spectacle *and*   6, 23
woman *and*   20–1, 23, 25–6, 65–6,
   105, 111, 124, 178
Palmer, D. J.   181 n.6
Panek, Jennifer   78–9
Pattison, Keith   161, 166
Payne, Alina   154–6, 158
Peacock, Lucy   17, 118, 120–1, 124–9,
   138, 178
performance   1–7, 11–28
   activism *and*   13
   audience *and* (*see* audience)
   convention (*and* tradition)   16, 32,
   38, 40, 45–9, 53, 55–7, 61, 63, 89,
   97, 103, 116, 126, 130–1, 133, 139,
   149, 151, 158, 162–3, 169, 172,
   174, 177, 184 n.2, 185 n.13
   despair *and*   103, 111–14, 117,
   139, 178
   feminist (*see* feminism)
   madness *and* (*see* madness)
   martyrdom *and* (*see* martyr)
   rehearsal of rape *and* (*see* rape)
   salvation *and* (*see* salvation)
   space   3, 45, 144–6, 159, 162, 171–2
   spectacle *and* (*see* spectacle)
   suffering *and* (*see* suffering)
   theory (*and* studies)   3–4, 13–14,
   19–22, 25–6, 29, 42, 45
   trauma *and* (*see* trauma)
   witness *and* (*see* witness)
   virginity *and*   158
Peter, John   166, 185 n.13
Peters, Christine   75–6, 98, 100, 104,
   107–8
Peterson, Kaara   111, 199
Phelan, Peggy   2, 5, 12–18, 22–6,
   42, 61, 90, 180
   Mourning Sex   13
   Unmarked   13–6, 22, 24–6
Phillippy, Patricia   64, 106–7, 114,
   184 n.7
Phillips, Kim   8, 34
Phoenix Theatre (London)   144
Pit, The (London)   183 n.11
Polycarpou, Peter   52
Portman, Jamie   130, 184 n.9
Post, J. B.   8, 13, 34

psychoanalysis 13–14, 48, 90, 183 n.13
Oedipal complex 48, 181 n.9
talking cure 89–90
see also Josef Breuer; Sigmund Freud; Julia Kristeva; Jacques Lacan
psychological realism *see* realism

Ramsden, Timothy 186
rape 2, 3, 7–11, 25–36, 40, 48, 49, 53, 71, 141, 143, 145–50, 156, 163–5, 169–70, 178
  abduction *and* 8
  advice to victims 32, 36, 38, 181 n.4
  aftermath of 33, 41, 50, 52
  and consent 8–9, 34–5, 39, 48, 146, 148–9, 169, 185 n.2
  effacement of 8–9, 11, 17, 21, 45, 61, 180 n.1, 3, 181 n.5
  history of 1–8, 13, 15–32, 41, 164–5, 168, 178–9
  law 2, 8, 10, 34, 149, n.180 n.2
  performance of 30, 63
  problem of rape's effacement 8, 9, 11, 181 n.5
  property crime *and* 8, 11, 33, 35
  rehearsal of 9, 30, 40, 49
  script 29–30, 36, 38–42, 49, 58, 71, 101
  trauma syndrome 51, 181 n.11
  victim of 3, 8–9, 11, 26–45, 49–51, 59, 61, 71–2, 144, 146, 148–50, 162–3
  virginity *and* 8, 157–8, 165, 170, 180 n.3
Ravenscroft, Edward 43
Rayburn, Corey 40
Rayner, Alice 5–7, 15–16, 22, 28, 40, 98, 102, 107, 119–20, 128, 180
realism 32, 59, 87–9, 126, 129–31, 183 n.13
  psychological realism 89, 176, 183 n.13
  see also naturalism
reception desire *see* audience
Rees, Ronan 124, 129

Reeves, Saskia 86–7, 89–90, 92–3, 96, 102, 133, 168, 178
Reid, Robert 130
Rhys, Matthew 58
Ridley, Bishop Nicolas 109–10
Ridout, Nicolas 162, 171, 173
Ritter, Sonia 51–6, 59, 86, 178
Roach, Joseph 2, 3, 5, 13, 180 n.4
Roberts, Anna 180 n.1, 181 n.5
Robertson, Karen 181 n.7
Robinson, Karen 128–9
Román, David 2, 13, 18, 28, 180 n.4
Rose, Mary Beth 77, 101, 104, 111, 183 n.6, 184 n.1
Rossini, Manuela 78–9,
Rowe, Katherine 181 n.6
Rowley, William 11, 141, 144, 151, 156, 164, 167, 174, 180 n.3
  see The Changeling
Royal National Theatre (London) 27, 95, 103, 130, 132, 136, 139, 183 n.14;
  Lyttelton theatre 95, 130, 137
Royal Shakespeare Company (London) 26, 32, 50, 67, 85, 87, 172
  Swan Theatre 54–5, 181 n.10

salvation, 10–11, 65–7, 73, 75, 77–80, 83–7, 93–7, 100–1, 106–7, 112, 116, 118
  narratives 67, 95, 110
  spectacle *and* (*see* spectacle)
  script 66–7, 93, 106–7, 112, 116
  performance of 3, 10, 26, 63, 66–7, 84, 98–101, 106–9, 112, 117, 134, 184 n.8
Scarry, Elaine 20, 23
Schneider, Rebecca 1–2, 4–6, 12–15, 41, 180 n.4
Scott, Adam 136
Scowcroft, Philippa 129, 184 n.9
Serlio, Sebastiano 154–6
sex 8–9, 11–12, 19, 27, 31, 33–5, 38, 61, 99, 104, 141–60, 142, 163, 168, 174–5, 180 n.3, 181 n.9, 182 n.14, 184 n.2
  female sexuality 144–5, 153
  forced sex (*see* rape)

Index   211

sex – *continued*
  sexual abuse, 163, 165 (*see also* rape)
  sexual choice 111, 148
  sexual desire 31, 35, 104, 114, 131, 142, 156–7, 163
  sexual innocence 9, 34–5, 38, 147, 181 n.3
  sexual martyr 111
  sexualized body 34, 90, 153
Shakespeare, William 1–2, 5, 11, 19, 26, 28, 30–2, 43–4, 57, 130
  Hamlet 5, 30, 38, 43
  Othello 22, 78, 176
  Taming of the Shrew 79
  see *Titus Andronicus*
Sheffield, Graham 172
Shevtsova, Maria 88
Shuger, Deborah Kuller 110
Shuttleworth, Ian 173
Simon, Roger 17–18, 23, 28, 117, 136, 138, 142, 177
Simpson, Herbert 184 n.9
Smart, Carol 180 n.1
Smith, Gary 129, 184 n.9
Soja, Edward 145
Sommerville, Margaret R. 9
Spears, Britney 60, 182 n.14
spectacle 2, 6, 9–11, 17, 19, 23, 27, 29, 38, 40, 44, 46, 50–1, 60, 64–5, 78, 87, 91–2, 99–100, 103–4, 109, 111–12, 117, 159, 162, 167, 173, 184 n.8
Spellman, Laurence 164
Spencer, Charles 54, 173, 185 n.13, 186 n.8
Stage Beauty (Richard Eyre) 176–7
Stanislavski, Constantin 177, 183 n.13
  Stanislavkian 88
Stimpson, Catharine 8, 181 n.5
Stockholder, Kay 185 n.2
Stratford Shakespeare Festival (Stratford, ON, Canada) 4, 27, 103, 117, 121, 124, 128–30, 136, 185 n.10
  Tom Patterson theatre 4–5, 117–19, 140
Stubbes, Katherine 63–6, 73, 78, 96, 106–7, 182 n.1–2

Stubbes, Phillip 63–7, 106–7, 182 n.1
suffering 2, 8–11, 17, 20–3, 25, 30–3, 36–7, 39, 46–8, 50–3, 56, 59, 62–5, 67, 71–2, 74, 76–7, 83–4, 86, 89–90, 95–9, 101, 103–4, 106–14, 120, 124, 128–9, 133–5, 138, 141, 144–5, 150, 167, 178, 162
  effacement of 101
  patient suffering 74, 113
  self-effacement *and* 44, 93
Sugimura, N. K. 185 n.2
Sumpter, Donald 51

Tapper, Zoe 176
Taylor, Diana 3, 11–13, 15, 18, 29, 102, 180 n.4
Taymor, Julie 26, 32, 56–61, 179, 182 n.13
Telford, Antony 185 n.2
Thompson, Mark 130, 137–9
Titus Andronicus 11, 26, 29, 31–2, 39, 42–62, 86, 129, 148, 151, 178, 181 n.5–7
  see also William Shakespeare
  see also Julie Taymor
  see also Deborah Warner
Titus 26, 56–60, 181–2
  see also Julie Taymor
Tompkins, Joanne 145, 150–60,
torture 5, 20, 22, 131, 134, 137, 139, 148
  see also violence
trauma 7, 9, 10, 13–14, 17–18, 30, 39, 42, 46–60, 82, 101, 108, 117, 119–20, 126, 127, 137, 142–3, 163, 165, 167–8, 175, 181 n.11
  sexual (*see* rape)
Tricomi, Albert H. 181 n.6

victim *see* rape
  see violence
violence *and* advice 10, 32, 36, 38, 71–3, 78–80, 107, 181. n.4
  effacement of 8, 9, 11, 17, 21, 25, 45, 61, 66, 85, 101–2, 116, 142, 146, 180 n.1, 3, 181 n.5
  and execution 99–100, 104, 106, 108–9, 111, 115–16, 126, 133–9

victim *see* rape – *continued*
  and fasting 78, 84, 183 n.10
  and feminism (*see* feminism)
  and the gaze 4, 7, 19–20, 25, 42, 57, 160, 180 n.5
  and invisibility (*see* invisibility)
  and pain (*see* pain)
  and pleasure 19–21, 57, 137, 147, 177–8, 184 n.5
  and reasonable correction 7, 9–11, 26, 65–71, 99
  representation of 3, 28, 141
  sexual 8–9, 27–63, 141, 143, 146, 149, 163, 165, 182 n.13 (*see* sex; *see* rape)
  and spectacle (*see* spectacle)
  and terror 21–2, 41, 48, 57–9, 61, 93, 98, 148, 163, 170, 175–6 (*see also* Kubiak)
  and touch 114–20, 122, 143, 165, 169, 184 n.8
  victim of 3, 8, 45, 85, 104, 131–3, 141, 144, 148, 150, 157, 160, 162, 181 n.4
  virginity *and* 8, 148, 181 n.3
  and witness (*see* witness)
Vitruvius, 153–5

Walker, Garthine 9, 35, 183 n.10
Wall, Wendy 105–7
Warner, Deborah 26, 32, 50–6, 59, 86, 129, 178, 179, 182
  see Titus Andronicus
Wayne, Valerie 75
Webster, John 2, 5, 11, 27, 65, 98–103, 111–16, 123–4, 126, 130–1, 136, 161, 180 n.2, 184 n.2
  Appius and Virginia 180 n.2
  see The Duchess of Malfi
Wentworth, Michael 78
Wentworth, Scott 124–5, 127, 134

Whately, William (*A Bride-Bush, Or A Direction for Married Persons*) 10, 26, 71–4, 66, 71–2, 74, 77–80, 83, 93, 133, 183 n.7
Whigham, Frank 104–5, 111
Wigley, Mark 152–3, 155, 157–8
Wiles, David 147
Wilkinson, Tom 176
Williams, Carolyn 8, 181 n.5
Williams, Linda 19–20
Williams, Olivia 143, 160–70, 178, 186 n.6
Wispé, Lauren 127
witness 3, 7, 10, 12–19, 23–33, 37, 39, 41–2, 45, 47, 50, 53–4, 56, 60–5, 67, 78, 85–7, 91, 93–103, 109–10, 115, 117, 119–31, 135–7, 139, 141, 150–1, 162, 171, 178, 185 n.12
  audience *and* 7, 13, 19, 23, 41, 50, 53, 56, 61, 97, 129, 184 n.2 (*see also* audience)
  engaged spectator *and* 28, 53 (*see also* spectator)
  eyewitness 17, 102
  feminist spectator *and* 179 (*see also* feminism)
  model 103, 129, 121
  onstage 86, 120
  performance *and* 3, 12–19, 22, 26, 180 n.4
  testimony *and* 18, 40, 64, 94
Wolf, Matt 136, 185 n.13
Wolskam, Aleksandra 92
Woman Killed with Kindness, A 11, 26, 66–7, 78–98, 100, 131, 168, 178, 183 n.9, 183 n.11, 184 n.9
  see also Thomas Heywood
  see also Katie Mitchell
Woodbridge, Linda 182 n.4
Wynne-Davies, Marion 181 n.7

Zimmerman, Susan, 38, 100, 108